PRAISE FOR DECISION SUPE

Eric Torkia and Bill Klimack have masterfully distilled decades of experience into a guide that is both accessible and profoundly insightful. This book is not just a manual for analysts but a blueprint for anyone looking to navigate the complex landscape of modern decision-making.

In a world where data is king, "Decision Superhero" empowers its readers to become the architects of their own destiny, leveraging the principles of decision science to drive impactful outcomes. Whether you're a seasoned professional or just starting your journey, this book will equip you with the tools and mindset needed to make better, faster, and more informed decisions.

With its blend of theory, practical application, and real-world examples, "Decision Superhero" stands out as an essential read for anyone looking to elevate their decision-making prowess. Eric and Bill have created a resource that is sure to become a staple on the bookshelf of every data-driven leader. This is the book that will turn good analysts into great decision-makers.

Jordan Goldmeier,
Author of *Becoming a Data Head* and *Data Smart*

Decision Superhero by Eric Torkia and Bill Klimack is a compelling guide to mastering decision science, crafted with clarity and depth. This book invites readers to explore the art and science of making better choices, blending real-world examples, historical anecdotes, and practical insights to create a rich learning experience. From operations research during WWII to contemporary decision analysis, the authors present a cohesive narrative that demonstrates the evolution of this critical field. The early chapters offer fascinating stories of decision-making challenges, such as Abraham Wald's work on survivorship bias, grounding readers in the timeless relevance of decision science.

The authors emphasize the significance of framing decisions and addresses the integration of data science with decision science - clarifying how data can support decision models, showcasing the complementary nature of these two fields. For professionals in leadership roles, there's a dedicated chapter on transforming organizations into decision-oriented firms that covers effective business transformation techniques for embedding decision science as a core competency. With its comprehensive approach and engaging style, Decision Superhero is an essential read for those looking to hone their decision-making skills.

Douglas W. Hubbard,
Author of How to Measure Anything: Finding the Value of Intangibles in Business and The Failure of Risk Management: Why It's Broken and How to Fix It

i

Decision Superhero provides an entertaining romp through history, commerce, science, and most importantly the treacherous middle ground between purely analytical work and model-free high-level decision making. Unafraid to blend insights from famous statisticians and Arnold Schwarzenegger in the same breath, Torkia and Klimack provide spirited advocacy for the notion that our mental models should often be replaced or augmented by more explicit, conscious modeling and careful analysis; that this can lead to surprising results and drive organizational value; and that it can often be accomplished with fewer prerequisites than we might be inclined to assume.

Peter Cotton,
Author of Microprediction: Buildiung an Open AI Network and,
Co-Founder and Chief Scientific Officer at Crunch Labs.

Decision Superhero by Eric Torkia and Bill Klimack is a standout guide in the field of decision science, penned by a pioneer whose expertise illuminates every chapter. They masterfully bridge the gap between complex theoretical concepts and practical business applications, making decision science accessible to data professionals of varying technical skill levels. Through historical anecdotes and real-world examples, they illustrate the impactful role of decision science across various industries and its pairing with data science to optimize business strategies.

The book is particularly beneficial for those accustomed to Excel, providing insights on how to leverage more sophisticated analytical techniques in the age of AI. Torkia and Klimack's clear and engaging writing ensures that readers can easily translate decision science principles into effective strategies within their organizations. Decision Superhero is highly recommended for anyone looking to deepen their understanding of strategic decision-making for business.

George Mount,
Author of Advancing into Analytics and Modern Data Analytics in Excel

Torkia and Klimack have successfully tied together high-level theories to real life anecdotes, making the concepts not only practical but actionable. The content being light-hearted and effervescent, making a topic that is often somewhat daunting for some, extremely accessible. A perfect tool kit for any manager or consultant needing to navigate successfully big and small decisions alike.

Sandra Abi-Rashed,
VP Business Growth & Development at Digilant and Founder of Mentoro

Decision Superhero

DRIVING INFORMED DECISION-MAKING WITH PROBABILITY, EXPLAINABLE MODELS, AND DECISION SCIENCE

Vol. 1 in the Decision Superhero Series

Eric Torkia and William K. Klimack

Technics Publications
SEDONA, ARIZONA

115 Linda Vista

Sedona, AZ 86336 USA

https://www.TechnicsPub.com

Edited by Jamie Hoberman

Cover design by Lorena Molinari & Eric Torkia

First Printing 2024

Copyright © 2024 by Eric Torkia and William K. Klimack

ISBN, print ed.   9781634626828 (paperback)

ISBN, print ed.   9781634626859 (hardcover)

ISBN, Kindle ed.   9781634626835

ISBN, PDF ed.   9781634626842

Library of Congress Control Number: 2024943270

I dedicate this book to Sam Torkia, my loving father and decision superhero prime who trained, encouraged, and fostered my professional development leading to this book.

I also dedicate this book to all the hard-working analysts being asked to do superhero things under impossible circumstances and succeeding. This book is by them and for them.

About the Authors

Eric Torkia holds a Master's degree in Management Information Systems and brings a wealth of experience in supporting strategic decision-making across various industries. As the executive partner at Technology Partnerz Ltd., he leads a team dedicated to delivering advanced decision science solutions focused on forecasting, simulation, and optimization. Eric advises senior leadership on business and analytics strategies, has trained hundreds of business analysts in predictive analytics, and spearheads risk analysis initiatives in finance and operations. He is also the lead developer of MCHammer.jl, a Julia package that streamlines Monte Carlo simulations. Deeply involved in the fields of decision science and business analytics, Eric balances his professional pursuits with the tranquility provided by his cats, affectionately known as his "corner assistants."

William K. Klimack has extensive experience applying operations research in energy and pharmaceutical companies. A retired U.S. Army infantry Colonel, he held faculty positions at West Point and the University of Oklahoma. He has a BS in chemical engineering from Lehigh University, a MS in applied mathematics from Johns Hopkins University, a MMAS from the US Army Command and Staff College, and a PhD in operations research from the US Air Force Institute of Technology. He is a Certified System Engineering Professional, a Professional Engineer, a Certified Analytics Professional, and a Fellow of the Society of Decision Professionals, the National Speleological Society, and the Explorers Club.

Acknowledgments

The Hall Of Decision Superheros

There are many people whose insights and support have made this book possible. I would like to acknowledge them here and thank them for their time, insights, and humor (because these guys are quite funny).

Sam L. Savage is a mentor and guru on all things simulation. In addition to many enlightening conversations on cutting-edge analytics, I have had the honor to work with Sam at PG&E as well as contribute ideas and present at the annual conference for probabilitymanagement.org, his non-profit dedicated to improving decision-making through a better understanding of probability. He is the godfather of vector-based simulation and how it can be used to federate simulation models across the organization (and beyond). Thank you for your support in writing this book and putting together the Foreword.

William McKibbin, PhD has been a resource and mentor since the near beginning of my career in decision science. He coached me on risk and probability along with how to teach them. Many of the foundational concepts presented in the book result from my early conversations with him way back when. Furthermore, he is a modern thinker and borderline futurist who, many times, has pointed me toward trends and ideas years in advance. Thanks Bill!

Robert D. Brown III is a dear friend and colleague who is the decision scientist's decision scientist, whose book "Business Case Analysis with R - Simulation Tutorials to Support Complex Business Decisions" inspired me to write Decision Superhero. We collaborated together on the "Need for Speed 2019" whitepaper comparing simulation speed and accuracy among the various Monte-Carlo alternatives. Over the years, we have shared and debated biases, probability, and expert elicitation. My vocabulary and intellect have both grown accordingly.

Bill Hill is a dearly departed friend that I must thank posthumously for being a mutant salesman who refused to take no for an answer. He is the one that could crash through a brick wall like the Kool-Aid Man, all covered in lime and mortar and look damn fine doing it. I first met Bill over the phone when he called me in 2006 to demo a Monte-Carlo tool called Crystal Ball. I tried to get rid of him, saying I did not need no Crystal Reports! Fortunately, being the tenacious guy he was, he convinced me to hear him out. I drank the Kool-Aid and have never looked at the world the same since. RIP Bill and thank you.

Eric Wainright is a long-time friend who is responsible for bringing Monte-Carlo simulation to Excel way back in 1991. When Oracle acquired Crystal Ball in 2008, I was introduced to Eric and we collaborated on designing features and enhancements to his seminal tool. This work inspired the creation of MCHammer.jl. Crystal Ball taught me analytics and Eric taught me some very important caveats regarding simulation and good software design. His contributions are considerable but unsung, so it is a privilege to highlight them here.

Egan Picken is a friend and client who took my quantitative risk analysis training using Oracle Crystal Ball years earlier. Later, we worked together to develop advanced option simulation and valuation tools in Julia based on their library of Excel models. I am most grateful as this project laid much of the technical groundwork for this book.

Peter Cotton is a cutting-edge author who developed micropredictions.org, a platform that has challenged the M6 competition for results using a crowdsourced approach to AI. He also happens to be a genius quant as well as a funny and relatable guy that made learning about new advanced portfolio optimization methods easier. Unfortunately, due to page count, I could not get into the many ways we can improve forecasting using his methods. That is for my next book. Nevertheless, his advice and insights helped make the book better. If you are not familiar with his work, you should look him up.

Contents at a Glance

Contents

CHAPTER 3

SuperPower: Developing a Firm that Thinks Differently

CHAPTER 11

SuperPower: Calculating The Value of Perfect Information _____ 253

REFERENCE

Figures

Tables

Icons and symbols

In this book we are accompanied by 3 superheroes, tasked with making sure you get the most out of the book.

The "Mind Power" Hero

 This superhero represents the power of critical thinking and deep reflection. The "Mind Power" hero ensures that readers pause and absorb the wisdom before moving on, helping them make the most of the intellectual challenges presented in the book.

The "Best Practice Guardian"

This superhero is the protector of efficient workflows and expert strategies. Whenever this icon appears, readers can trust that they are being guided toward a tried-and-true method that will enhance their decision-making skills. The "Best Practice Guardian" helps readers take practical steps in their learning journey, equipping them with tools they can rely on in real-world scenarios.

The "Hands-On Hero"

 The superhero pencil marks the moment when readers can get their hands dirty by following along with example files. When this symbol appears, it signals that downloadable resources are available, allowing readers to apply what they've learned in real-time. This hero ensures readers stay engaged with interactive exercises and reinforces their learning through hands-on application, making the journey from knowledge to practice seamless and exciting.

Foreword

I was exposed to some of the ideas in this book prenatally in 1944 when my father, Leonard Jimmie Savage, was at the Statistical Research Group at Columbia University. He was applying mathematics to war operations with Abraham Wald, a Decision Superhero described in Chapter 1. The field of Operations Research developed in both the US and England during WWII and, accelerated by the development of computers after the war, spawned the decision sciences of today.

Aspects of this field, like riding a bicycle, can't be learned from a book, but require an experiential component. So, naturally some books on this subject teach only the parts that can be taught from a book. I find these too academic as they can leave the reader "educated," but unable to "ride the bike." Torkia and Klimack are seasoned practitioners with experience in a wide variety of real world applications in both the military and industry. So, this book, paradoxically, also teaches things you can't learn from a book.

But how?

Software models play the role of "decision bicycles," that provide an experiential interpretation of the written word. This connects the seat of the intellect to the seat of the pants in an approach I call Limbic Analytics.

And importantly, in the more technically focused books, two types of software are used: Excel, which is ubiquitous, and Julia, which is a fully programmable language.

But wait, isn't one of these environments better than the other? Why teach both? This is like asking which is better, pencil and paper or the printing press? The authors make it clear that "Developing models is an iterative process." This is the only series of books on modeling I have seen that suggests, correctly, that you can start with Excel to rough things out and create prototypes, then move on to a formal development platform, such as Julia, when you are ready to go to press.

Reading all three Decision Superhero books at once is not a casual exercise, but you don't need to read them all at once. Learning is nonlinear. Think of this book series as a large museum, like the Louvre, which you will keep coming back to. You probably already have Excel but may not have Julia. If so, you can pass over the Julia examples and return to them as you want to develop more production-oriented applications.

Chapter 1 provides a great historical context for the power of decision science. Then look at the roadmap to get the lay of the land. Wherever you go in the museum, you will find the exhibits to be clear and entertaining.

Enjoy.

Sam L. Savage

Author of "The Flaw of Averages"
President of probabilitymanagement.org

Introduction

It's Friday, 3 pm, and your boss asks you for urgent analysis and insight, what do you do? Why you?

In the movie *Office Space*, the manager is an expert at dumping these kinds of assignments on analysts' desks, prompting a game of cat and mouse to avoid working on the weekend. Of course, this is a caricature, but there is some truth to being in situations where we need to make big decisions over the weekend with analysis needed by Monday morning. *We all wished we had a secret recipe to get through the analysis quickly and in time, and that's why we wrote this book series.*

Our first book, the one you have in hand, is focused on the ideas, thinkers, mathematics, and psychology of making decisions. This book gives you many tools to tackle big decisions and rank opportunities. The second book takes a more technical turn by teaching computational thinking and modeling using Excel and Julia. Here, we learn how these tools are similar and different and how to use them together to maximize their effectiveness. The third book is for those who want to take their modeling to the next level, showing advanced methods such as correlation, time-series fitting, simulation, and numeric optimization for when scenario analysis and figuring out what's best is the name of the game.

This book series did not happen overnight. This is the fruit of 20 years of conversations with smart people, lots of assignments, and many scars. I was very fortunate that fellow decision superhero, Bill Klimack, was able to further refine the precepts and history of the main theme of the Decision Superhero series: Decision Science. Bill started out in the military but eventually transitioned to applying his operations research and decision analysis skills as a consultant and an executive responsible for developing and managing decision quality in several oil and gas companies where decision science is already considered a core competency. This rich experience in corporate life and consulting was a fountain of best practices that helped us shape this book. Bill coauthored the chapters with me, ensuring we covered all the most important elements for the budding decision superhero while ensuring certain ideas and themes were present in other parts of the book. I am most grateful for his help and his major contribution to the theory and practice of decision science.

Decision science is an interdisciplinary field that uses various quantitative methods to enhance decision-making processes. However, it's not just about numbers and algorithms; it's about understanding the stories these numbers tell and harnessing this understanding to drive actionable insights. As you embark on this journey, you'll learn the skills and superpowers of the modern Decision Superhero, who combines strategy, data, and tools to make informed choices, shaping the destinies of businesses and organizations.

Decision modeling is central to decision science and, by natural extension, this book. Almost every chapter touches on a practical dimension of decision modeling because it provides a structured framework to test assumptions, understand variables, and predict outcomes. This book focuses on the foundational ideas of decision science, processes, workflows, axioms, and rules for making quantitative decisions. But for a Decision Superhero to be successful, it is not just a question of having a bunch of cool mathematical tools but of having a clear understanding of what it takes to make high-stakes decisions as a facilitator and as an empathetic and sentient human.

Like all professionals, a decision superhero should learn effective soft skills in time, but who has the patience for that? This is why we have put a special focus on what it means to be a decision superhero, such as the mental attitudes, soft skills, and dispositions that make collaborating on decisions an effective and appreciated exercise by those involved—including you!

If you are a leader seeking to transform how decisions are made and supported with insight, then we have included an entire chapter to the ideas and tools necessary to catalyze a transformation towards an organization that thinks differently, especially about the central importance of decision science as a competency. We cover the structures and strategy approach a decision-oriented firm takes, including the Dynamic Enterprise Alignment model, which proposes a new way to conceive alignment and performance. None of this would be complete without some business transformation techniques to ensure the adoption of new ways of doing things.

Data science has been all the rage in the last few years, and its focus on machine learning has provided the tools and methodologies necessary to advance AI to today's state. Though AI and Data Science are different in their vocations, the success stories of data-driven decision-making in various industries highlight the potential of AI, driving adoption and popularity.

These data superheroes are essential in getting data and assisting decision superheroes in developing insights that management can use to chart a course to better performance. The math is the same, but the approaches and focus are different. Chapter 4 highlights the similarities and differences between these two disciplines and how they can work together to achieve maximum effect.

Decisions are models and models are decisions, which is the title of our most philosophical chapter. Because models crystallize one's thinking, they are very important instruments in the work of decision scientists. We go through the different model types and how we would characterize our work. For example, few know the difference between numeric and analytical models—we cover that with many others. The goal is to give the decision scientist a holistic view of the tools available to build toy models of the world with which to test ideas.

With data science and decision science models being central to corporate decision-making, it is important to know when to use which. People often reach for what they know, and sometimes that decision is worth revisiting. Insight costs money and sometimes automation is the way to reduce costs and get results more often, translating into an improved signal-to-noise ratio. Picking the right tools and approach for the job is tantamount, especially from a *bang-for-the-buck* point of view.

Making this work successful in an organization often requires a roadmap that defines the endpoint with steps in between. This chapter proposes a series of destinations with a workflow to get you there. We propose a master process for decision science, with each element broken out into workflows that will allow you to design business analysis projects and decision science processes that tick all the boxes. Once again, you will find a series of best practices for creating wins that also encourage management to further fund the development of decision science within the firm.

Decisions, the core of the discussion here, have special and universal properties that allow for design and evaluation. We cover all the elements of a decision, including those that happen outside our heads. Decisions are often subjects that touch many people. Sometimes, many people are involved in designing a decision, and that's when decision framing becomes critical. So, as you would expect, we put in a section on *how to frame like a champ.*

Chapters 9-12 cover calculating probability and using it to make more informed decisions considering risk and uncertainty. We explore the classic decision theory methods, which lay the groundwork for all other analytical decision-making. Many important concepts in modeling decisions are discussed, along with interpreting the result using Savage's axioms, utility theory, decision rules, and risk attitudes. Though these methods are simple and useful, they are equally necessary when we analyze the results of a Monte Carlo simulation.

We include two hands-on case studies. One to calculate the value of perfect information and a hands-on case to decide under complete uncertainty. The methods are simple and proven, and we include all the Excel files that work through the ideas. Understanding the classic decision theory approach sets the stage for what comes next.

Modeling is a vast topic, and the practices and theorems vary from discipline to discipline. Given the centrality of modeling to the success of a decision superhero, in Books 2 and 3, we shall delve deep into various modeling techniques that apply universally, from deterministic simulations to predicting a myriad of futures using probabilistic methodologies.

As we said from the beginning, this is a journey and a destination. Let the journey begin.

ONE

Origin Story for the Modern Decision Superhero

If you think about it, getting what you want or achieving what you want always leads to making some sort of decision. Want to improve your business? Want to improve your life? The key is to make better decisions more often and with better outcomes. This is why we want to help you become a *Decision Superhero*!

Fortunately, Decision Superheroes have Decision Science. In essence, decision science is a mix of soft skills (interviewing, elicitation, countering bias, etc.) and hard analytical approaches (math, statistics, modeling, machine learning, etc.), which allows for a more thorough understanding of the dynamics of a problem, as well as the levers to improve its outcome using an iterative, science-based approach. A science-based approach simply means that you are starting with a question and trying to deduce all the elements needed to estimate an answer and help with its understanding.

While decision science goes back a long way, its power has not been lost in the modern world *by those who remember it*. Decision science originated in operations research problems during the Second World War. Using probability and statistics to better understand the world and make decisions has origins going back another 400 years. However, decision science, as we know it today, has its roots in decision theory, a discipline that evolved quickly in the 1950s and 1960s. Some of history's most notable thinkers and mathematicians are the superheroes of this story. People like Eratosthenes, Daniel Bernoulli, Pierre-Simon Laplace, Jimmie and Sam Savage, John Von Neumann, and Abraham Wald.

In this introduction to decision science, we will discuss its early beginnings in history to demonstrate the curiosity and creativity that solved big problems and emerged as its own discipline. This book will prove that in a complex world, there are techniques to solve problems, gain insight, and translate inspiration into action.

The stories and anecdotes on the history of decision science through the following pages will focus on some tales that show the power of thinking mathematically and making better decisions.

With so many decisions, maybe it should be a science

Every day, from when you wake up to when you go to sleep, you're faced with numerous decisions. Some are minor, like deciding what to wear (unless you have a special event) or choosing between eggs and French toast for breakfast. Others are routine yet essential to starting your day right, such as picking the best route to work or deciding between coffee and tea. Then, there are the choices that are even more basic, like whether to use paper or plastic bags at the grocery store or put sugar in your coffee. These decisions might seem small, but they can add up quickly. For example, reducing calories is a big deal when someone is trying to lose weight. If you elect not to take sugar in your coffee or forgo that piece of cake, over time, the *cumulative power* of those decisions impacts your waistline!

On the flip side, you're also occasionally faced with bigger, more impactful decisions. These are the kind that can change your life or affect others significantly. For instance, deciding on a selling price for your house, accepting a new job or a different pay structure (like stock options), making investment choices (such as between bonds or stocks), figuring out the best way to balance the risk in your investment portfolio, considering starting or investing in a new business, or even making a lifelong commitment like getting married. These decisions are full of uncertainty and can significantly impact your life and those around you.

Making these weightier decisions can be daunting, mainly because they're often surrounded by *uncertainty and ambiguity*. When faced with such decisions, it's crucial to take a step back, slow things down, and analyze the situation carefully. This thoughtful approach allows you to consider all your options thoroughly, ensuring you make the best possible choice. Rushing into decisions without this analysis can lead to regrettable outcomes. Probably summarized best by Franz Kafka in this seminal quote, *"There art two cardinal sins from which all others spring: Impatience and Laziness."*

This wisdom highlights the importance of taking the time to make well-considered decisions and not shying away from the effort required to analyze complex choices thoroughly. This is where decision science comes in. It offers a structured and objective way to tackle decision-making, especially when the stakes are high. By applying decision science principles, you can make more informed, high-quality decisions over time, navigating life's uncertainties with greater confidence and peace of mind.

Ralph Keeney, a respected Professor Emeritus at Duke University's Fuqua School of Business, shares in his book, *"Give Yourself a Nudge"* a simple yet powerful idea: decisions are key to enhancing our lives and businesses. One of his standout ideas is value-focused thinking, a concept about improving how we make decisions.

Given that we're always making decisions, big and small, it stands to reason that consistently making better choices can significantly improve our lives. Despite this, history and various studies show we often struggle to make good decisions, especially the big ones. Benjamin Franklin, for instance, suggested a method of listing the pros and cons of a decision and comparing them to make a choice clearer—a technique that points to an early understanding of decision-making challenges.

Keeney put his theories to the test with an experiment involving MBA students. Having invested time and money into their education, these students were asked to list their reasons for pursuing their degree. On average, they came up with 6.4 objectives. However, after engaging in activities designed to deepen their thinking, they realized they had about four more goals they hadn't initially considered! This shows how important it is to fully explore our reasons behind a decision to avoid making choices with incomplete information.

Typically, we make decisions as a response to problems, identifying and evaluating different solutions until we find the most suitable one. This approach, which Keeney calls alternative-focused thinking, is by nature reactive. Instead, he proposes a shift towards value-focused thinking that starts with identifying what truly matters to us—**our values and objectives**—and then keeping an eye out for ways to achieve these. Instead of seeing decisions as mere solutions to problems, we begin to view them as chances to proactively pursue our goals, thus turning decisions into opportunities rather than decisions being forced upon us by circumstances.

Now that we have made the case that making better decisions is good business, we can get started. However, before going off and droning on about decision science and using the term indiscriminately over a few hundred pages, we feel it is important to outline the origins of the discipline and provide a unified working definition.

A science by any other name...

William Shakespeare famously wrote, "*A rose by any other name would smell as sweet.*" Decision science has not always been known as such. In the past, you may have done decision science without knowing it, as unknowingly expressed by French playwright Molière's Monsieur Jourdain in the Bourgeois Gentleman: "*By my faith! For more than forty years I have been speaking prose without knowing nothing about it, and I am much obliged to you for having taught me that.*"

The realms of decision analysis, management science, and operations research are intricately linked, each contributing distinct perspectives and methodologies towards the overarching goal of improving decision-making processes. Their development over time reflects a progressive evolution toward what is now often collectively termed decision science.

The main goal of this book is to offer readers a systematic and rational methodology for problem-solving and decision-making. While not novel, this methodology has been a cornerstone of scientific inquiry and practical application for decades. Originating in the 1940s, it was initially employed by management scientists and operations researchers to address military challenges of the era.

Our focus will extend this approach to address contemporary management issues in business, government, academia, and similar institutions.

In this context, we use "Decision Science" to denote the application of the operations research and management sciences approach within the realm of decision-making. Thus, the terms "operations research" and "management science" may be used almost interchangeably with decision science throughout this book.

At its core, operations research is about understanding the system surrounding a decision-making problem. It involves identifying controllable variables, determining the objective, recognizing un-

controllable factors, and understanding the interplay between different elements within the system.

Management science, often used interchangeably with operations
research in some contexts, focuses on applying analytical methods
to solve managerial and strategic problems. It encompasses a broad
range of techniques, such as linear programming, simulation, and
statistical analysis to optimize organizational performance, resource
allocation, and strategic planning.

The roots of management science can be traced back to the early 20th
century, with significant growth occurring after World War II. The
need to solve complex logistical and operational problems during
the war using operations research methods led to the formalization
of management science as a discipline. Its post-war application to
business problems marked the beginning of its widespread adoption
in the corporate world.

By the 1960s, many large organizations, such as GE and GM, started adopting *decision analysis*. Scholars like Ronald A. Howard and
Howard Raiffa saw decision analysis as a response to the need for a
structured approach to making complex decisions, particularly in
the context of business and engineering.

Decision analysis is a systematic, quantitative, and visual approach
to making high-quality decisions under uncertainty. It uses a variety
of tools, including decision trees, payoff tables, and sensitivity analysis, to evaluate the outcomes of different decisions. Its primary focus
is on individual decision-making processes, emphasizing the clarity
of preferences, trade-offs, and the impact of uncertainty on decision
outcomes.

The evolution from operations research and management science to
decision analysis and eventually to decision science reflects a broad-
ening of focus from purely operational and managerial issues to a
more comprehensive view of decision-making.

> *While operations research and management science provided the mathematical and analytical foundation, decision analysis (theory) introduced a structured framework for dealing with uncertainty and preferences in individual and organizational decisions.*

The convergence of these fields into decision science signifies a recognition of the complexity and interconnectedness of modern decision-making. Decision science integrates the quantitative rigor of operations research, the strategic focus of management science, and the clarity and structure of decision analysis. It embraces a multidisciplinary approach, drawing on economics, psychology, and systems engineering to address the multifaceted nature of decisions in a rapidly changing world.

This evolution reflects a continuous effort to refine and expand the tools and methodologies available for decision-making. As challenges become more complex and the stakes higher, the principles of decision science offer a beacon for navigating uncertainty, optimizing outcomes, and achieving strategic goals.

The story of the man and the boat

Imagine a man, stranded in a lifeboat in the middle of the ocean, his cruise ship vanished beneath the waves. In his desperation, he looks to the heavens and pleads, "God, please save me!"

Not long after, a helicopter buzzes overhead, lowering a ladder towards him. The pilot shouts, "Grab on!" But the man waves them off, saying, "No need, God's got my back."

A few hours later, a rescue boat motors up alongside him. The crew tosses over a rope and urges, "Come on, we'll pull you to safety!" Yet again, the man declines, "Save your efforts; I'm waiting for a divine rescue."

As the sun sets, a pair of dolphins appear, nudging the boat and signaling toward land. "Thanks, but no thanks," the man tells the dolphins. "I'm holding out for a miracle."

Eventually, the inevitable happens—the man doesn't survive. Standing before God, he's frustrated: "I trusted you! Why didn't you come to my rescue?" God, with a sigh, responds, "Because you're an idiot. I sent a helicopter, a boat, and even a pair of dolphins. What more were you waiting for?"

Always remember to be open to opportunity because the answer to your prayers may not be obvious or look like what you are expecting.

A modern definition of decision science

Until now, we have explored the evolution that resulted in what is now called decision science. At a high level, the term **Decision Science**[1] refers to the collection of qualitative and quantitative techniques and a science-based iterative approach to maximizing **decision quality**.[2] More formally, we propose the following unifying definition:

> *Decision Science is an iterative approach to maximizing decision quality that provides a clear enough view [prediction] of a strategy's [unbiased] reality to allow decision-makers to make the appropriate strategic, management, or control decisions necessary to achieve its goals.*

We can view the process as encountering a question about what to do, developing a hypothesis that guides gathering information, analyzing it, and making the decision (see *Figure 1*).

Question ➡ Hypothesis ➡ Research / Data ➡ Analysis ➡ **Decision**

Figure 1: Basic decision research process.

In its simplest form, *a decision is an action-based answer* to a question. In some situations, the question defines the decision made, but sometimes it's the other way around. In the context of this process, we think the answer will be a hypothesis. This is a necessary anchor or starting point. The beauty is that the starting point can be wrong, but the process should eventually tell you so and perhaps suggest an alternative that has not been considered.

---

[1] We note that our definition is not universally accepted as there are many different uses of the same terms. For example, James Wetherbe says *that decision science is an alternate name for management science.*

[2] The probability that a decision, if based on the analysis, will deliver as expected.

All authors and researchers agree that decision science must lead to decision-making and, by extension, action.

By now, it should be clear that there is a long history around math and decision-making. Here are some classic tales of pernicious decision problems where past decision heroes used intuition, creativity, and math to save lives.

Decision science and the art of war

Modern decision science, with its rich tapestry woven from numerous mathematical and modeling innovations, owes much to the crucible of war. These contributions, born out of necessity, have profoundly shaped the field. Let's embark on a brief historical journey to explore the origins of key ideas, problems, and challenges within decision science, tracing their lineage back to wartime innovations. This exploration also sets the stage for understanding concepts like "victory disease" and "survivor bias," revealing how past triumphs and challenges influence decision-making strategies today.

Victory disease

If not knowing the end goal wasn't bad enough, people often have poor intuition when making decisions. In military theory, there is a concept called *Victory Disease*. The idea is based on the common historical observation that the victor in the last war lost the first battle in the next war (some may recognize the pattern in sports playoffs). The principal theory is that the loser is more open to self-examination and innovation than the victor; thus, the loser is usually better positioned to win at the start of the next conflict.

Some can argue that the French suffered from Victory Disease after World War I. Following the German surrender, the Allied nations were not complacent. France built the *Maginot Line,* an expensive static defense system, to protect itself from German attack. After the start of World War II, the Germans attacked France and the Low Countries. In a period of weeks, they drove the Allied forces back. The British Army on the continent would have been entirely lost if not for the incredible evacuation from Dunkirk, which saved many British soldiers and some French. Still, their equipment was lost and would take time to replace. The British then had a limited ability to fight the Germans, and their army needed to rebuild. The Royal Navy, the largest in the world, was spread across the globe, while Germany was primarily a land power and less vulnerable to sea power. To solve this unexpected inequality of force, the British turned to the *Royal Air Force (RAF)*.

Between the wars, military air power visionaries argued that strategic air power was the way to avoid the brutal trench warfare of World War I. Based on air combat experience in the first war, many adopted the slogan, *The bomber will always get through*. In 1939, RAF Bomber Command flew missions against German naval targets. In the first named air battle of the war, German fighters intercepted and mauled British bombers over Heligoland Bight. The British concluded their air power assumptions were incorrect, and the RAF, as organized, was not an effective force against the Germans. Whatever their initial perspective, the British were not suffering from Victory Disease after the withdrawal from France. They were open to innovation and ideas from anywhere. Another example from World War II is when university professors were consulted for ideas related to protecting military aircraft.

Discovering survivorship bias

Abraham Wald, one of the first Decision Superheroes, was a Jewish Hungarian mathematician who fled Austria due to the growing Nazi influence and was asked while working in the United States to look at aircraft damage on the planes that made it back, not always in good shape either.

At the time, antiaircraft gunners aimed at a plane's center of mass, with the sighting system compensating for target velocity, wind effects, and altitude and range impacts on the munition trajectories. The RAF looked at aircraft damaged after combat missions to consider where to add armor. Initial conclusions were to up-armor the wings and fuselage. Wald noted that the data was obtained from *aircraft that returned* while the real solution lay with the missing data from the aircraft *that did not make it back!* This is also known as **survivorship bias**, which is a significant way of not framing the problem correctly. Fortunately, Wald's observation and work led to protecting specific areas of the plane that would lead to a crash if damaged, such as the engines. Although overstated, the story has been passed down that the initial guidance to armor certain areas was rescinded, and the instructions became to armor everything else—the reverse of the first set of instructions. **The initial assessment had the right mental model but misunderstood the data.**

Post-World War II, the success of the wartime operations research techniques was noted and adopted by those returning to the civilian sector. Russell Ackoff, a professor at the Wharton School, was a proponent of applying operations research to the private sector. Ackoff was perhaps the first to codify the general approach of the operations researcher.

On the soft side of decision science, intellectual and interpersonal skills are necessary to design, construct, and decide based on a practical understanding of the information and the context under which it is given. Unfortunately, humans tend to flash-analyze information and use biases and generalizations to make split-second decisions. In some cases, this is exactly what we need when faced with clear and present danger, while in others, it may lead to making hasty and wrong decisions. Because the world is full of biases that cloud our analyses and lead to wrong decisions, we will briefly examine an interesting aspect of behavioral decision-making.

A solution when intuition leads you astray

The aircraft armor example shows that *truly understanding* the problem can be the key to finding the best solution. Unfortunately, we often think we understand issues even when we do not. In his book, *Thinking Fast and Slow*, Nobel Laureate Daniel Kahneman writes that you can describe how we think as having two competing systems in our heads. **System 1** is an automatic system that runs cheaply and quickly leaps to conclusions (we jokingly call this *low-calorie logic*). System 1 works well for routine things by applying routine mental models. If you have ever driven home and realized you do not remember the drive, System 1 was engaged. **System 2** requires mental effort and takes over when concentrating on a complex task. Answering 1+1 is handled by System 1, while counting down from 100 in increments of 4 is a System 2 task.

System 1 is continuously operating and will leap to answers. While System 2 can override System 1, System 2 may not notice the issue. Look at Figure 2: Müller-Lyer illusion, a famous optical illusion. The lower horizontal line looks longer. Now, measure them. They are of equal length. When you look at the figure now, System 2 knows the lines are equal because of the act of measuring. But System 1 is still fooled, and the illusion persists. Besides optical illusions, we can have cognitive illusions, which drive us to leap to the obvious and incorrect solution. This is a form of bias. The key takeaway is that System 2 can override System 1 when the problem is spotted, but if it is unnoticed, which means you are not looking, System 1 is in charge.l leap to answers. While System 2 can override System 1, System 2 may not notice the issue. Look at Figure 2: Müller-Lyer illusion, a famous optical illusion. The lower horizontal line looks longer. Now, measure them. They are of equal length. When you look at the figure now, System 2 knows the lines are equal because of the act of measuring. But System 1 is still fooled, and the illusion persists. Besides optical illusions, we can have cognitive illusions, which drive us to leap to the obvious and incorrect solution. This is a form of **bias**. The key takeaway is that System 2 can override System 1 when the problem is spotted, but if it is unnoticed, which means you are not looking, System 1 is in charge.

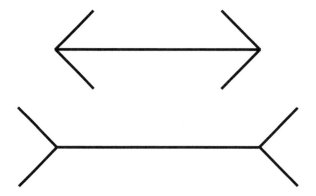

Figure 2: Müller-Lyer illusion

However, even when the situation is perfectly understood, and we engage System 2, or slow thinking, we can make poor decisions. The long-running television game show *Let's Make a Deal* is based on the choices made by a contestant:

1. A contestant is shown information about a valuable prize hidden behind one of three doors and told that the other doors conceal unattractive gifts: goats.

2. The player then selects a door.

3. Next, the host, who knows what is behind each door, opens one of the two unselected doors to reveal a goat prize.

4. The contestant is now asked whether they would like to switch to the other unopened door.

Does switching doors improve the probability of winning? Intuition says it does not matter, but decision science provides a mathematical answer. Each door initially had a one-third probability of having the valuable prize. Now, the probability is equal between the two remaining doors; therefore, switching should have no effect, right? Unfortunately, the current door retains a one-third probability of winning while the other door has a two-thirds probability. Counter-intuitively, switching to the other door provides the highest probability of success. Weird, right? This is known as the **Monty Hall problem**, named after the show's original host.

On the surface, the answer seems incorrect. The strength of how incorrect it seems was demonstrated when a column by Marilyn vos Savant appeared in a 1990 issue of *Parade magazine*. Her column provided the correct answer, and 10,000 readers responded with almost universal disagreement. Eventually, using a computer simulation largely settled the issue in her favor. We have reproduced this simulation in Excel using simple `IF()` statements to demonstrate.

	A	B	C	D	E	F	G	H
1	Excel Only Solution to Monty Hall							
2	*10,000 trials. Hit F9 to rerun simulation							
3				Stay and Win?	Switch and Win?			
4			Win Probabilit	33.38%	66.62%			
5								
6	Car	Your Pick	Monty Opens	Stay and Win?	Switch and Win?			
7	A	A	C	Yes!	No!			
8	B	A	C	No!	Yes!			
9	A	A	C	Yes!	No!			
10	A	A	C	Yes!	No!			
11	C	A	B	No!	Yes!			
12	B	A	C	No!	Yes!			
13	B	A	C	No!	Yes!			
14	B	A	C	No!	Yes!			
15	A	A	C	Yes!	No!			
10002	A	A	C	Yes!	No!			
10003	B	A	C	No!	Yes!			
10004	C	A	B	No!	Yes!			
10005	A	A	C	Yes!	No!			
10006	A	A	C	Yes!	No!			
10007								
10008								
10009								
10010		Door	Cutoff					
10011		A	0					
10012		B	1					
10013		C	2					
10014								
10015		=IF(A2=Sheet2!$A$2,IF(RAND()<0.5,"B","C"),IF(A2=Sheet2!$A$3,"C","B"))						
10016								
10017		* This is a technique where we assign 3 discrete states using the RAND()						
10018								
10019								

Figure 3: Excel simulation of the Monty Hall problem using conditional logic.

In the preceding figure, the probabilities resulting from a 10,000-trial Monte-Carlo simulation show a 67% chance of winning if you switch versus staying with the original pick. The cognitive illusion was identified by empirically measuring the number of times switching pays off.

Obviously, this mathematical illusion was understood by someone; otherwise, System 1 is continuously operating and will leap to an-

swers. While System 2 can override System 1, System 2 may not notice the issue. Look at Figure 2: Müller-Lyer illusion, a famous optical illusion. The lower horizontal line looks longer. Now, measure
them. They are of equal length. When you look at the figure now,
System 2 knows the lines are equal because of the act of measuring.
But System 1 is still fooled, and the illusion persists. Besides optical
illusions, we can have cognitive illusions, which drive us to leap to
the obvious and incorrect solution. This is a form of bias. The key
takeaway is that System 2 can override System 1 when the problem
is spotted, but if it is unnoticed, which means you are not looking,
System 1 is in charge.e no money. The Monty Hall problem con-
fuses even System 2, but mathematical analysis and modeling provide clarity. The Müller-Lyer illusion and the Monty Hall problem
demonstrate that by using heuristics, your mind can play tricks on
you. In our first example, which line, if any, is longer, and secondly,
which door maximizes the probability you do not wind up with the
goat? Most decisions are not as clear-cut and require more work to
structure and organize before you can do any accurate modeling or
programming.

Decision science made the monorail faster

Besides avoiding negative
consequences, approaching
decision-making correct-
ly can yield positive sur-
prises. Main Street USA is
the entrance to the Magic
Kingdom at Disney World
in Orlando, Florida, and
was designed to represent
a small town in Missouri,
where Disney grew up. The

Figure 4: The Walt Disney World monorail train.

street-level features cater to the shopping preferences of the guests.
Above the street level are windows bearing names of fictional businesses and their proprietors. Their names are real and placed there
to honor those who have made extraordinary contributions to the

Disney Company. One such tribute window honors Bruce Laval, another early Decision Superhero, who developed the FastPass system in the attraction and made other contributions for many years. Laval, an industrial engineer who had just finished an MBA program focusing on **operations research (OR),** attracted attention as a recent hire at Disney. Disney was experiencing capacity and wait time issues with the monorail train, an iconic entrance system into the Magic Kingdom. Visitors found that there was often a backup at the boarding station. Having visitors wait in a long line even before they got into the park proper was distressing to the park management. The prevailing wisdom and what appeared like an obvious solution was to add passenger capacity by purchasing an additional train.

Laval was asked to write up the justification for buying a new train. Using his OR education, he built a computer simulation to empirically study the problem and remain objective about the outcome. This is also one of the key benefits of modeling decisions using simulation and computers; it allows for objective analysis. His modeling and simulation provided counterintuitive results: *Adding one more train did not resolve the passenger wait issue. Removing one train did!* Many were left scratching their heads until further analysis of the results revealed that the futuristic experience of riding the monorail, one where no operators were on board, had unforeseen consequences. In more conventional systems, human operators would use their judgment to optimize traffic. The impact of using a fully automated driver-less experience was not realized until human intervention was removed from the system, and it underperformed design expectations when subjected to real-world conditions.

More specifically, Laval's analysis revealed that a safety aspect of the system was causing the backlog. If two trains came too close to each other, both would slow to a very reduced speed, thus causing a capacity problem under crowded conditions. Because boarding passengers would delay a train's departure, the next train would encroach into the safety buffer space. Both trains would then move at a reduced rate, causing all the following trains to go to the reduced rate, leading to a growing passenger queue that just compounded the problem.

Conversely, Laval's analysis recommended removing a train, which permitted the trains to avoid encroachment and operate at normal speeds, thus allowing Disney to avoid purchasing a costly and unnecessary train while still providing a solution that delivered the intended entry experience. Laval gathered the data and found the mental model was incorrect through the use of an objective computer model.

From dietary allowances to optimization

Economist George Stigler proposed a diet problem in 1943. He asked, *what would be the least expensive combination of 77 foods that would meet the government-recommended dietary allowances?* This was a linear programming problem and Stigler was able to estimate an answer using trial and error approaches. Stigler observed that the diet met the guidelines but was not very palatable. Several years later, George Dantzig developed the simplex algorithm and found the true optimum value, which was slightly better than Stigler's. This might have all been esoteric, but the idea played an important role in the Berlin Airlift. Early in the Cold War, the Soviet Union blockaded Western nation surface access to Berlin, which sat inside East Germany. Only aircraft could be used to provide food and fuel to Berlin's population. The German diet favors potatoes, but potatoes are heavy compared to their nutritional value. A form of the Stigler diet problem was used to provide nutritional food that fitted within the airlift capacity. (Aircrews flew over 200,000 sorties in about 11 months, successfully supplying Berlin, and the Soviets stopped the blockade.) Stigler went on to earn the Nobel Prize in Economic Sciences in 1982 for his work on the effects of regulation on markets.

The Traveling Salesman Problem

The Traveling Salesman Problem (TSP) is a classic problem where a person tours locations, visiting each only once, and returns to the trip's origin. The salesman wants to take the shortest route to save time and expense. Heuristics and exact algorithms are available to solve the problem, but they are computationally intensive. The problem has been considered for some time, certainly in the 1800s by Sir William Rowan Hamilton. Julia Robinson first used the TSP moniker in a 1949 RAND report.

The computational challenge is illustrated by a 1954 Newsweek magazine column that described the TSP with visits from Washington DC to a major city in each of the (then) 48 US states. The article states *three Rand mathematicians worked to produce a solution in only a few weeks* that would cover travel to 49 cities using the shortest possible route. The problem was such a challenge that the Proctor & Gamble company held a contest in 1962 offering $64,000 in prize money for the best solutions for the shortest round trip visiting 33 cities.

The traveling salesman algorithm is ubiquitous in today's world. Route planning websites are available on the Internet. Logistics systems calculate different routes based on cost. Distance and other metrics that go beyond identifying the shortest route. Package delivery drivers scan the packages loaded on their vans and then run an optimization program to get their day's route. This is quite an advancement from the days when drivers were told to manually plan routes to avoid left turns. The package delivery story illustrates the rapid progress in improving computational ability.

As we saw in the previous examples, the ability to make better decisions generally translates into surprising and positive benefits. But, as in anything, if we want to reap the rewards, we must make the investment. Sadly, in many organizations, it's not always easy for an analyst to get in front of the decision-maker with their analysis and data in hand, and make an influential contribution. Structuring your analysis using a decision science approach will allow you to make even more compelling arguments, as well as be able to defend all that great code and data analysis you put into your report. At the end of the day, a good idea without execution is useless. That's why a decision superhero has to get their idea in front of the right people—otherwise, it's a waste of time.

How the ancients calculated the earth's circumference

Take the case of *Eratosthenes,* who, inspired by a simple observation of the sun, correctly predicted the circumference of the planet using geometry. Let's quickly explore how Eratosthenes, one of history's earliest decision superheroes, pulled this off.

By happenstance, Eratosthenes noticed while reading a papyrus scroll (from the great library of Alexandria) that in Syene, at noon on the summer solstice, the sun shines directly down a well, illuminating the water at the bottom. This meant the sun was directly overhead and would cast **no shadows**. This was very interesting because in his home in Alexandria, at noon during the summer solstice, he had previously observed the main tower *did* cast a shadow.

Perplexed, Eratosthenes started pondering how this could be. Could it be that the Earth is curved and not flat? He knew that if the Earth was indeed curved, the difference in the angle of the Sun's rays between Alexandria and Syene would show shadows of different lengths and provide a clue to the Earth's circumference.

Eratosthenes realized that if the Earth was flat, the angle of the Sun's rays at both locations

Figure 4b: Meta AI's take on Eratosthenes explaining his theory

would be the same. However, since the Sun's rays are parallel and the Earth is curved, the difference in the angle of the Sun's rays between the two locations corresponds to a fraction of the Earth's total circumference.

Being the decision superhero that he was, he devised a simple but effective strategy to get an answer:

1. **Setting Up the Experiment**: Eratosthenes dispatched a friend to Syene to pace out the distance between Alexandria and Syene (about 800 kilometers). Once there, his friend planted a rod to measure the length of the shadow at noon on the summer solstice.

2. **Making Observations:** In Syene, the vertical rod casts a shadow, and by measuring its length and the angle of the Sun's rays, Eratosthenes' friend found that the angle was about 7.2 degrees.

3. **Calculating the Circumference:** Knowing the angle difference (7.2 degrees) and the distance between Alexandria and Syene (about 800 kilometers), Eratosthenes used simple geometry and proportion to estimate the Earth's circumference.

He knew that 360 degrees make up a full circle, so the fraction of the Earth's circumference corresponding to the 7.2-degree angle could be calculated as or 1/50 . Therefore, if the distance between Alexandria and Syene (800 kilometers) represented of the Earth's circumference, the entire circumference would be 800 kilometers x 50 = 40,000 kilometers.

His model results were not bettered until an exploration party physically measured the distance 2,000 years later! It turned out that Eratosthenes was only off by 2-3%, which is not bad for the time. It is clearly a case of validation in the real world taking a really long time.

As a more contemporary example, take an experimental design for a plane that started out as a drawing only to see it fly in the sky a few years later. Imagine correctly gauging the demand for a product that has never been sold before. Both examples started as predictions and were either proved right or wrong by reality.

Key takeaways

- The first step in becoming a great decision superhero is to understand both the core definition and history of decision science. The second step is *knowing* that the more decisions you make that go your way, the more profitable or successful you will be in the long run.

- Since the early days of the modern era, decision-making has evolved into a science driven by a series of historical successes and, more importantly, failures. We looked at several examples in history where creative thinking and not taking the obvious solution led to innovation and saving lives.

- We also took some time to consider the importance of correctly framing a decision, as well as where decision science can contribute to the successful definition of a model that you can code or build out into a spreadsheet. These concepts will serve as the foundation for ensuring that the models we develop answer the correct questions.

In the next chapter, we will look at some of the skill sets, competencies, and personal attributes that would make you a badass decision superhero.

Two

The Decision Superhero (a.k.a. Scientist)

B ecoming a decision superhero is a quest to develop skills, competencies, and behaviors that translate into tangible results. In the following pages, we will explore the ideas, convictions, and principles that make the difference between just being another analyst and being a decision superhero.

And what if you want to develop more decision superheroes? Good news! We will explore the various soft and hard skills and technical competencies a decision superhero should have, as well as the ethical concerns for wielding such power.[3] Time to get your mask and tights, the journey begins.

---

[3] Books 2 and 3 of the Decision Superhero series cover how to comprehensively build advanced decision models using Excel and Julia.

Meet the decision superhero

*Decision Scientists bring unbiased clarity to
decisions through modeling and common sense.
A **Decision Superhero** (a.k.a. **Decision Scientist**)
is a person who assists management or stakeholders with applying a structured modeling
approach using all the available and appropriate
data and math, including probability, to provide
clarity and an unbiased perspective on the drivers impacting the desired outcome, leading to
more informed and robust decisions.*

We use the term **Decision Superhero** as a tribute to the titanic efforts analysts and decision-makers need to solve problems and answer questions about the future, often with little or no time.

A decision scientist provides objective analysis from a third party
perspective who has the vocation of being unbiased. If a decision
scientist is able to avert or minimize the impact of a bad decision
(usually figuring in the many thousands or millions of dollars), a
good decision scientist should pay for their salary many times over
in a year. Sometimes, these individuals can come across as the pro-
verbial pain in the ass because they are constantly acting as a devil's
advocate, further making their point with data and established modeling practices. This can be frustrating to some, even though arguing
the point of a decision and supporting the chosen option is good
practice. It works for the US Army, and it worked for some brilliant
historical figures as well.

Be more right than wrong

"It is better to be vaguely right than exactly wrong."
Carveath Read

Any time the value of an analysis or a recommendation outweighs the cost of doing so, you should have a decision superhero (quantitative risk and decision analyst). This is especially compelling considering that, according to the Standish Group's 25-year study of IT projects, *almost 70% of projects fail to deliver as expected or not at all.* That means that some 20-30% of projects get canned before completion while almost 50% hobble across the finish line, costing too much, taking too long, or with a considerably reduced scope.

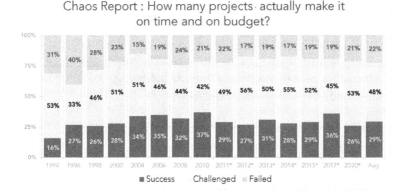

Figure 5: Chaos report 1994-2020. How many projects actually make it on time and on budget?

A decision superhero will help management understand the factors driving success and failure by giving early risk information.

The applied practice of decision science is similar to having a mathematical soothsayer.[4] You elect to pay for the time and effort of a

---

[4] "a person who predicts the future by magical, intuitive, or more rational means": prognosticator - Merriam-Webster.

decision analyst to improve your odds of betting on the right horse. You know the odds are in your favor, but because there is uncertainty involved, there is no guarantee that your selected horse will win. While you might still lose that bet, this approach will produce the best results in picking horses in multiple races.

Decision superheroes have secret identities

Regardless of how your analysts are allocated, they sit in the layer between the people who generate data and the business needs of those who make decisions. Sometimes, you have Decision Superheroes in the firm, but they go under different names depending on either the function or line of business. Some names include *quantitative risk analyst, decision analyst, business analyst, financial analyst, strategic planner, analytics professional, and so on.* **The key defining feature of decision scientists, regardless of title, is that their analysis is intended to produce a decision recommendation.** But why not use an AI if making a recommendation is all you need? Context. Contrary to most existing systems, a decision scientist can and should be able to tell you why they are making a recommendation, linking context and environmental information.

There are other types of analysts, but their focus is on designing detailed solutions (business processes, automation, new technology, etc.) rather than identifying options for action (as presented in the following figure):

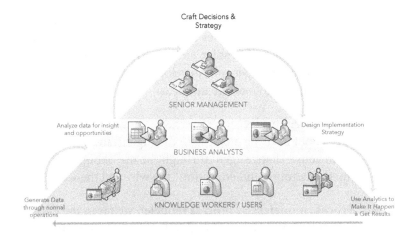

Figure 6: How do analysts fit in the business?

Decision superheroes may assume many roles, each requiring different competencies, frequently to support decision-making within an organization. Here are some of the main costumes a decision superhero may have to wear:

- **Facilitator:** A decision superhero must be effective all along the continuum from soft interpersonal skills to hard technical skills. At the soft skills end, a decision superhero must be able to facilitate meetings and workshops. The framing process is a facilitated workshop and is key to the success of the process.

- **Business coach**: Usually, the decision superhero is the person most experienced at thinking about *decision quality (DQ)*. The decision superheroes should provide guidance in structuring and analyzing decisions. Additionally, decision superheroes should work individually with project team members and others in an organization to help them understand the value of decision science and how to apply its tenets to better decision-making.

- **Project leader:** The decision superhero may be a member of the project team supporting the decision-making, but they may also be the project leader. A project leader may also be a project manager, which means both planning and managing project activities in addition to quality assurance and producing deliverables. In these types of roles, soft skills take on a bigger role, developing internal networks and the ability to manage upwards.

- **Instructor:** The decision superhero brings unique knowledge about decision-making to the organization and should be an active change agent through the use of both training (methods/tools) and education (the why and the what).

- **Process modeler:** By process modeler, we mean the decision superhero may need to develop models to replicate the functioning of portions of the business or process. *For example, in support of oil and gas exploration and production decisions, earth scientists and petroleum engineers generally build a simulation to represent how the hydrocarbon reservoir will perform once production begins.*

- **Economic modeler:** A decision superhero also must understand that financial and economic aspects affect the decision. A simple example is correctly interpreting the effects of the time value of money on cash flows when calculating net present value. *Economic aspects can be quite complex for some contractual agreements and tax regimes, and when correctly analyzed, may suggest a course of action contrary to conventional wisdom.*

- **Quality assurance:** Quality assurance ensures models are useful because they were verified and validated. Typically, verification checks that the model performs as intended, while validation ensures that the verified model can be used to answer the questions of concern. Decision superheroes should be prepared to support the organization by quality controlling their own work and, as needed, ensuring that the work of others is also error-free and appropriate for use.

- **Independent project review:** A key tenet of decision science is that it should be fair and independent. At times, decision superheroes may have to push back against projects or the readiness to support a particular business unit when it's not beneficial to the overall organization.

Decision science plays a crucial role not just in choosing a course of action, but also in shaping how that decision is executed, since the nature of the decisions vary and the required analytical techniques are particular to the problem being solved. Decision superheroes can have different roles and entry points into the process. They offer valuable analysis and insights to those tasked with carrying out the chosen strategy. For example, a company planning to build a new warehouse to improve its delivery efficiency. The decision on the warehouse's capacity and design specifics are integral parts of the decision-making process. Overlooking these aspects could lead to the warehouse incurring higher costs and underperforming, ultimately undermining performance rather than enhancing it.

In addition to the plurality of roles that a decision superhero can assume, the way a firm is organized to handle decisions and their implementation can have a huge impact on how effectively these are accomplished.

Decision science sales challenges

An important point to remember is that the insights created by analytics (decision and data science) are often tied to individuals (decision superheroes) and not processes. Though more difficult and challenging on an organizational change level, this phenomenon makes developing broader and more integrated analysis **necessary** for them (decision scientists) to extend the value to other areas of the business.

Another important consideration is that decision science organizations rarely have the responsibility and accountability for the creation of insight as well as the execution of that insight. Selling the value of decision science should not be some fluffy or lofty goal for managers interested in sustained results—it is a necessary criterion for success.

The Zen mind of the decision superhero

"When we have the true sense of the Tao, of the real knowledge and wisdom, we will be able to make the right decisions in our lives."
— Mantak Chia

Decisions are not an easy business. A decision superhero will often be in complex and challenging situations where Tai Chi on eggs is the best if not the only strategy for success. In the following sections, we shall explore some important affirmations, which are Zen mantras you should keep in mind when things go nuts, to bring balance to your discussions and direction to your project—in other words, how to stay sane in a crazy situation.

Affirmation #1: A decision scientist is objective

"All models are wrong, some models are useful." — George Box

The preceding quote is the most direct advice we can give to understand the necessity of intellectual honesty, a propensity for clarity and precision, and healthy skepticism in modeling. Remember it well because a great analyst is humble (about results) and strives to maintain objectivity on the data and the methods. The relationship between an analyst's humility and accuracy is proportional. Nate Silver, in his 2012 book, *The Signal and the Noise: Why Most Predictions Fail—but Some Don't*, makes this point convincingly by exploring Isaiah Berlin's story of the Fox and the Hedgehog—hedgehogs, who view the world through the lens of a single defining idea, and foxes, who draw on a wide variety of experiences and for whom the world cannot be boiled down to a single idea. Silver's point is that a good analyst is a fox because it keeps its mind and eyes open to new information and ideas. We agree! A fox is constantly scheming and looking for new ways to get the chicken, while the hedgehog is content with what it can see in front of it.

Affirmation #2: I NEED less data than I think, and I HAVE more data than I need (to make a decision)

Doug Hubbard, in his seminal book, *How to Measure Anything*, makes a very compelling point about how to correctly handle the objection of not having enough data. I had the chance to meet Doug at a conference held by Dr. Sam Savage on Probability Management back in 2016. As you would expect, Doug is a very sharp but funny guy. I remember him saying, "If astrology were an accurate forecasting method, I would use it!" Who can argue with that?

He posits that a solution for making a decision is seldom impossible because a solution probably exists somewhere else and that the amount of data required to carve out an estimate is usually smaller than what we initially thought. Thus, everything (useful, that is) can be measured or estimated. If we omit this, giving up too soon, or losing focus on solving the problem because we are trying to get a perfect answer (paralysis by analysis) become potential outcomes.

Doug summarizes his approach to estimating business problems as follows:

- *Your problem is not as UNIQUE as you think*

- *You have MORE DATA than you think*

- *You need LESS DATA than you think*

- *There is A USEFUL MEASUREMENT TOOL THAT IS EASIER than you think*

Consider who has the data you will need and how to obtain access. Data may be objective—available as recognizable data or subjective, based on opinion. Subjective data, data from people, can be extremely important. Technical experts must be available for elicitation sessions to gather the data. Additional IT resources may be needed for the data or intensive modeling.

Affirmation #3: Not all decisions work out, not even the best-made ones

The most important competency for a decision scientist is an understanding of, and passion for, achieving decision quality (DQ). The idea of DQ is to have a forward-looking metric that provides assurance that you are making a good decision. Because almost any decision involves uncertainty, making a good decision has a higher probability of leading to a good outcome. However, uncertainty makes no guarantee that a good decision will produce a good result. You could make a good decision, but a cruel chance may produce an unwanted outcome.

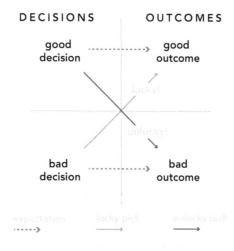

Figure 7: Decision versus outcome.

Affirmation #4: Remember that decisions impact others

Martin Yate is the author of a highly successful series of books on navigating the job search process. A key point of his is that at the core of every job is the task of identifying, preventing, or solving problems. While identification may not involve any prevention, solution, mitigation, or other actions, it certainly requires decisions. Like decisions in our personal lives, some decisions are straightforward, while others require careful consideration. Personal decisions can affect others. Assuming you take Yate's job search advice and (hopefully) get multiple attractive employment offers, it may also require making trade-offs that affect others. If an offer requires you to relocate, that will affect family members, sometimes in unforeseen or adverse ways.

In the same way that personal decisions can impact others, this is almost always the case with business decisions, as they often touch multiple stakeholders. The normal definition of a stakeholder is anyone who affects or is affected by a decision. Currently, Walmart sits atop the Fortune 500 list and has over 265 million customers each week. Walmart's collective decisions affect its workforce, which stands at 2.2 million employees, as well as its suppliers and customers. With such large sets of stakeholders, decisions are sometimes complicated. Keep in mind the following: First, we need to understand who the decision-maker is. Second, how the decision-maker and the stakeholders will be impacted, and third, decision context.

Affirmation #5: Keep the focus on the prize

The title decision scientist suggests the centrality of decisions to the role. The idea behind decision science is to be decision-focused. This means two things: tailoring the work to the decision and being alert to recognize decision opportunities. Tailoring the work to the decision means doing work that will help make the decision, and only that, work.

Science is the devoted study of the world through observation and experimentation. In the exploration and production portion of the oil and natural gas industry, earth scientists play key roles in understanding the potential hydrocarbon reservoirs deep under the surface. As scientists, they would always like to gain more information, but if that information is not deemed to affect a decision, it is not valuable from a business standpoint. Furthermore, assuming there is some value, if the value is less than the cost of acquiring more data, it is not economically valid either. Defining and clarifying a decision's context has proven useful in making such analyses. For this purpose, the US Army War College has adopted the acronym **VUCA** to characterize the context in which to make decisions:

- **Volatility** is the degree to which the situation is characterized by sudden and significant change.

- **Uncertainty** refers to the fact that even when we know what the drivers are, the outcomes we will see are unknown.

- **Complexity** describes the nature of the context itself and the degree of intricacy and interconnectedness of factors bearing on the decision.

- **Ambiguity** is the degree of difficulty in discerning the structure of the situation.

Being observant of decision opportunities is the other aspect of being decision-focused. Know your values or objectives and be vigilant for opportunities to pursue them. As an individual, a decision superhero helps foster decision focus by contributing to the organization's culture and providing the tools and insight for managers and leaders to do the same thing.

Affirmation # 6: Know thy domain

Decision superheroes need to have enough knowledge to function within the application area in which they work. Ideally, decision superheroes have industry experience. Industries are complex and typically multi-functional teams are brought together. Individuals on the team bring unique knowledge, as well as their domain knowledge. The result is that what is obvious to one person on a team may not be apparent to another team member.

> *No one's an expert in everything, so decision scientists do not have to be the complete master of the domain knowledge.*

However, they must be conversant enough with the concepts and vocabulary to facilitate meetings and engage with the project team. Decision scientists must have a concept of how the system will work. At the core, there are two key points. Decision superheroes must be able to communicate in the domain vocabulary and must provide reassurance that they will be able to provide effective insight to support decision-making. If this is an issue, decision superheroes should serve as process experts and be paired with a domain expert to facilitate effective management of the project's decision-making.

Ethics: The truth and nothing but the truth

All professionals are expected to behave ethically, but sometimes situations are complicated and interests are varied, leading to fuzzy ethics. Clearly, some things are legal but not ethical, further muddying the waters. Here are a few critical considerations when pondering action or responding to requests.

Who do you serve?

In the 1972 Oscar-winning film *The Godfather*, we were introduced to the Corleone mafia family. The head of the family, Don Corleone (the Godfather), played by Marlon Brando, made all the decisions but was advised by his *consigliere* (a word for a trusted advisor in Italian), played by Robert Duval. Much can be learned about Machiavelli and his management practices in this story, but the one thing that made it easy for Robert Duval over most decision superheroes is that he knew who his boss was, while in other cases, it may not be so clear. For this reason, it is very important to understand whose interests we serve and who is responsible for paying us. Loyalty is a very big component of ethics, but misplaced loyalty can result in jail. Sometimes, because of history or the length of a relationship, we stay involved in an unhealthy or dangerous situation long past its due date. That being said, being sensitive to the needs and objectives of your patron is incredibly important and will serve you later on.

You are both advocate and devil's advocate

Once you are clear on who you represent and advocate for, you can seek opportunities to maximize gain or minimize the probability of some bad event. But given that there are other ideas, opinions, and positions that will differ from that of your patron, your job as a decision superhero is to proactively identify the problems/objections for all the stakeholders involved in the decision you are working through. By being the devil's advocate, we are striving to look at both sides of the equation to identify things that will prevent the smooth rollout of our project. If we can identify and mitigate objections through active listening and maybe a little empathy, then we can rest assured that our project will move forward much faster within the organization.

What is being truthful?

"The amount of energy needed to refute bullshit is an order of magnitude bigger than [that needed] to produce it." – Alberto Brandolini

Beyond credibility issues, *"bullshit"* is a costly proposition for any organization and should be avoided, as evidenced by Brandolini's law[5]. For analysts, this can be dangerous territory because the only people that can recognize you are *"bullshitting"* with the numbers, intentionally or not, are those who know more about the topic than you. Those who know more can either be safety nets that will point out tunnel vision, or snipers who will shoot you down in a bloody and public way. Since there is no real way of knowing for sure who is in the room with you, humility, being truthful, and the ability to acknowledge a concern diplomatically and publicly is almost always the way to go.

Being truthful as an analyst means telling it as it is, based on all the information you have on hand. You may be wrong, but at least you took *all the care in preparing the data and selecting the methods to ensure the best possible answer—that is your fiduciary[6] obligation.* As an analyst, some people may ask you to back into an answer (produce new-school BS), which can often happen in but is not limited to the field of financial engineering. It is quite human to want to stack the cards in our favor, but it is not the decision superhero's job to facilitate that. Backing into an answer, which is different from having an initial hypothesis for the results, happens when someone tells you the answer they want beforehand and you have to work a model to support that answer. *This is unethical and should be avoided at all costs.* In fact, if someone asks you to back into a number or you suspect they are asking you to, then the only correct response is

---

[5] First said on twitter by Alberto Brandolini, Jan 11, 2013. (https://twitter.com/ziobrando/status/289635060758507521?s=20).

[6] A fiduciary duty is a commitment to act in the best interests of another person or entity. Broadly speaking, a fiduciary duty is a duty of loyalty and a duty of care. That is, the fiduciary must act only in the best interests of a client or beneficiary. And the fiduciary must act diligently in those interests. (https://www.investopedia.com/ask/answers/042915/what-are-some-examples-fiduciary-duty.asp).

to confront them by saying, *"Are you asking me to back into a number?"* Over the years, I have resorted to this technique, and I can assure you nobody has ever responded with *"Yes."* That would be tantamount to saying, *"Yes, I am asking you to lie."* Mind you, had they, the engagement would have ended right there. It is better to walk away from a bad deal than to have it strapped around your neck like a noose for the rest of your career. Someone once mused that his grandfather gave the advice that you should never sacrifice your name for less than a million dollars... We agree but for inflationary reasons, we would add a few zeros.

The ethics ACID test

Probably the best and simplest advice when thinking about ethics is to check whether you would be comfortable with what you are about to do being highlighted on a news feed or the newspaper's front page. If not, you should carefully reconsider.

Developing your superpowers

> *"The one who says it cannot be done should not interrupt the one doing it."* – Arthur Block

Skill sets fall into two broad categories, which merit equal attention and care if a decision scientist wants to be equipped to meet the modern challenges presented to them.

Soft skills are desirable qualities that do not depend on acquired knowledge but rather on social and interpersonal skills (such as listening, empathy, common sense, right mental attitude, and so on). In contrast, **hard skills** focus on technical knowledge.

A core soft skill for a decision scientist is the ability to communicate effectively in writing and orally. The need to communicate is embedded into facilitation but especially comes into play when the decision-maker is updated. Communicating effectively takes advantage of, and accounts for, the other work in the decision-making process.

Business superpowers

What are the hard and soft skill sets that a decision scientist needs to be successful? Obviously, this is a spectrum, and no one can be expected to know all the stuff in Table 1. However, it is important to know more than less in this list.

TABLE 1: DECISION SCIENTIST HARD AND SOFT SKILLS

DIMENSION	SOFT SKILLS	HARD SKILLS
People	Interviewing	Social network analysis
	Facilitation	Organizational change management
	Expert elicitation	
	Active listening	Written communication
	Networking	Presentation skills
	Business communication	Data visualization
	Managing meetings (whether virtual/in person)	Graphic design (making impactful documents)
	Coaching and mentoring skills	HR and compensation model development
	Teaching skills	

DIMENSION	SOFT SKILLS	HARD SKILLS
Operations	Systems thinking	Process modeling
	Domain expertise	Discrete event simulation
	White boarding	Financial modeling
	Strategic planning	Project management
		Benefits management
		Business continuity management
		Quantitative risk methods
Technology	Ability to simplify	Total cost of ownership analysis
	Requirements gathering	Application portfolio optimization
	Keeping an eye on innovations	Options comparisons
	Ability to translate strategic objectives into analytics project options	Technology architecture
		Data management (DBs and other data technologies)

DIMENSION	SOFT SKILLS	HARD SKILLS
Business Alignment	Industry knowledge of alliances and collaborations	General business acumen
	Organizational awareness	The appropriate contractual, technical, and financial knowledge to support modeling
	Functional industry vocabulary and knowledge	Alliance management
		Network analysis
Continuous Improvement	Facilitation	Six sigma
	Collaboration	Statistical quality control
	Coaching	Lean methods
		System modeling

If you are just starting, picking business superpowers is the base for planning future learning or deciding which projects to take.

Once you pick a project, it will probably influence all future work because that is the experience you bring to each and every project. So don't do something you dislike because that may mean you will do a bunch of projects you don't like!

In addition to business superpowers (domain expertise), a decision superhero must be familiar with some essential analytical tools. Some can be done quickly in your head while talking or negotiating, and others require time, thought, and analysis.

Analytics superpowers

Decision superheroes come in many shapes and colors and, there-fore, proficiency in certain skill sets may be greater for some than others. One thing is for sure—an analyst must have analytical skills of some sort. Some may notice, through careful examination of the hard and soft skills of the decision superhero, that there is a fair amount of overlap between the skills required in data science and decision science.

TABLE 2: DECISION SUPERHERO HARD AND SOFT SKILLS.

SKILL SET	SOFT SKILLS	HARD SKILLS
Math	Knowledge of the OR and analytics skills discussed in this book	Probability
		Statistics
	How not to bore people to tears when presenting your idea	Hypothesis testing
		Math programming
		Algebra
		Graph theory
		The Markov decision process
Modeling	Building mental models	Building spreadsheets
	Whiteboarding	Visual modeling
	Influence diagrams	Programming in Julia
	Decision trees	
Data	Finding data	Exploratory data analysis
	Assessing which data source is correct if multiple sources exist	Eliciting unbiased data
	Data storytelling	Dashboarding with PowerBI, Tableau, or Spotfire
	Translating business requirements to data needs	
		Data cleansing
	Communicating data requirements to relevant stakeholders	SQL queries
		Data normalization

Quantitative methods in decision science

Quantitative techniques used in decision science include those from the fields of decision analysis, risk analysis, cost-benefit and cost-effectiveness analysis, optimization, simulation, behavioral decision theory, operations research, statistical inference, management control, cognitive and social psychology, data science, analytics, and computer science.

Different problems have different sets of tools and solutions for getting your answers. If you mismatch the method to the problem, you may end up with the wrong answer or take too much time to get to something meaningful. At this step of the process, we select the most appropriate analytical methods and where to source relevant data. Though decision science is not operations research, it borrows almost all its models and mathematical methods but makes the allowance that they can be applied to almost any decision. *The 2020 Institute of Industrial and Systems Engineers Industrial and Systems Engineering Body of Knowledge* identifies 14 categories of knowledge. One of these categories is **Operations Research and Analysis**, which contains the following:

- Linear Programming
- Nonlinear Programming
- Transportation Problems
- Linear Assignment Problems
- Network Flow and Optimization
- Deterministic Dynamic Programming
- Integer Programming

- Meta-heuristics
- Decision Analysis and Game Theory
- Modeling under Uncertainty (Stochastic Processes, Markov Chains, Stochastic Programming)
- Queuing Systems
- Simulation
- Systems Dynamics

The adoption of these methodologies was aided as computing power improved. This is illustrated nicely by the history of the TSP in Chapter 1.

Key takeaways

- Decision superheroes are hardworking analysts who often have to move mountains and defy the notion of time and space to meet impossible deadlines.

- Decision superheroes help their leadership and patrons to make better decisions by crunching the numbers, making sense of complex situations, and accounting for the psychology and biases of the stakeholders involved.

- Though we have seen that the decision superhero is a rare combination of intellect and wiles, we have also noticed that these special individuals can be found spread out and hidden in the far reaches of most organizations.

- By adopting a Zen and holistic approach to problem modeling and solving, the decision superhero always seeks clarity and reduces uncertainty enough to make a good decision and spearhead projects that deliver results quickly.

- Finally, we saw that a Decision Superhero should be deeply ethical and empathetic to the stakeholders they work with to be fully effective and credible.

Of course, Decision Superheroes are not alone in using data to advance organizational goals; in chapter 4, we will meet our acolyte, the Data Superhero, and discover how we can team up for even greater results!

THREE

SuperPower: Developing a Firm that Thinks Differently

D eveloping decision science as an organizational capabili- ty is both a journey and a destination. This chapter will cover the ideas and concepts necessary to allow you and the group you work in to hopefully embrace an old but new way of approaching decisions relating to allocating resources to maximum strategic and operational effect.

A natural relationship between decision science and resource-based strategy stems from both being focused on the optimal allocation of resources. One is concerned with analyzing the best way to do it, and the other focuses on making it happen. A firm that wants to have decision science as a core competency will also accept that it needs to have ways to influence how strategy is planned–this is why they work together so well.

The Dynamic Enterprise Alignment model allows for the clear integration of these frameworks and guides where decision science can make a positive impact.

We will also consider how decision superheroes navigate firms, drive change, support people making tough decisions, structure processes, and enlighten the masses (okay, maybe a little ambitious). Even in the most disciplined organization, knowing how to best organize and allocate your best resources (itself a decision science problem) is critical to getting results.

Consultants or develop skills internally?

For managers contemplating enhancing decision science capabilities within their organization, the dilemma often boils down to a fundamental "make or buy" decision. Opting for the "make" approach entails internally developing these capabilities, leveraging existing personnel and resources. Conversely, the "buy" strategy involves enlisting external resources, such as consulting services or academic expertise, to fulfill this need. This external reliance can be either a temporary measure to bolster internal capabilities or a long-term solution.

Mobilizing resources, encompassing both human and technological assets, is a significant aspect of project planning and organizational development. It necessitates careful consideration of the requisite cultural shifts and structural adjustments within the organization. This is because integrating decision science resources—whether sourced internally or externally, temporarily or permanently—presents unique challenges and opportunities.

Effective decision scientists often transcend their initial roles to become invaluable advisors to project teams and key decision-makers. Achieving this level of influence and integration is a gradual process that requires strategic planning and alignment with the organization's goals. The suitability of different approaches to developing decision science capabilities will vary depending on the existing organizational structure and culture. Regardless of the current configuration, it is crucial for firms to regularly evaluate their approach to ensure that it continues to provide optimal support for decision-making processes.

This ongoing assessment ensures that the organization remains agile and responsive to the evolving landscape of decision science, thereby maintaining or enhancing its competitive edge in the marketplace.

Decisions start with the end in mind

From a practical point of view, we can put decision science into a two-part system that focuses on *identifying/framing opportunities* and *selecting an optimal strategy* that delivers desired outcomes.

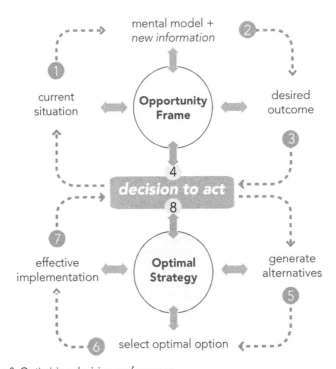

Figure 8: Optimizing decision performance

At the center of this approach is a **decision to act**. There are two decisions: one to decide if there is a decision to be made and another to identify an optimal decision.

Opportunity framing

Opportunity framing is a distinct and important process whose objective is to delineate the parameters of a decision, including if it's a decision worth making. I am often reminded of the old saying, *"You can't get there from here,"* when decisions do not have a clearly defined end-point or specific criteria for knowing it paid off. This creates a myopic view of where potential business opportunities can come from, and this is where decision science comes into play. The first step is to document where you are in terms of relevant internal Key Performance Indicators (KPIs), external benchmarks, and generally being aware of the trends affecting the context of the decision and, more broadly, the organization. Becoming good as an organization at framing decisions is not something that happens overnight. *With this in mind, you will notice that framing is covered in its various aspects at different points of the book, including an entire chapter dedicated to techniques to develop powerful decision frames that are useful in evaluating alternatives and ultimately making a decision.*

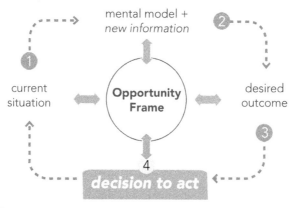

Figure 9: Opportunity Framing Cycle

In his 1977 book, "Arnold: Education of a Bodybuilder," Arnold

Schwarzenegger articulated a principle distinguishing sustainable improvement from short-term gains. He emphasized the importance of acknowledging and addressing one's weaknesses, contrasting this approach with that of other bodybuilders who might conceal their less developed areas. Schwarzenegger argued that real progress comes from the willingness to expose and work on these vulnerabilities rather than hide them. This philosophy underlines a broader principle applicable beyond bodybuilding: developing a robust decision process requires a clear recognition and focused effort on areas needing improvement. In essence, effective management and enhancement are contingent upon not just measuring performance but critically acknowledging current shortcomings as opportunities for growth.

Once you've identified things that would make sense to work on in the organization, you then need to set measurable objectives to assess whether your decision science initiatives are doing any good. The result of the first loop is twofold: *A decision for action and an objective where success needs to be objectively measured. Classic stuff.*

Optimal strategy selection

The decision that we need to do something should trigger asking, "What should be done?" The first step is to identify as many reasonable and perhaps unreasonable solution ideas that could improve or resolve the situation. Coming up with interesting and innovative alternatives to evaluate is where creativity comes into play.

There are many ways to go about designing creative solutions. One very effective method that also applies to framing problems is the Work-out process, which we cover in Chapter 8. Independently of how you define your courses of action (alternatives), once defined, they can be submitted to different states of nature (scenarios/trials) to identify the optimal option.

Sometimes, there is more than one solution to a problem. Therefore, a portfolio approach that balances multiple initiatives towards a greater objective is required. Nevertheless, options need to be weighted, compared, and analyzed for their ability to satisfy the requirements of the decision. Fortunately, Chapter 12 provides a robust framework for this, which is also applied when analyzing the results of simulation plus optimization models, which we will cover in books 2 and 3 in this series.

Execution is systematically cited in business literature as the most important criterion for success when comparing the various capabilities that truly successful organizations have. Executing effectively starts with setting up an objective, which we did in the optimal *strategy loop*. Of course, some solutions take several iterations before truly delivering the intended results. When in doubt, you just need to look at software or smartphones.

Because decisions must always result in action (initially to go forward or not), the implementation team takes over and makes a first go at realizing the decision.

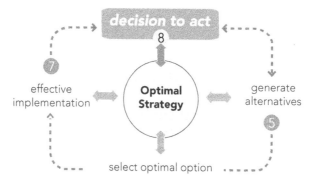

Figure 10: Solution Development Process Loop in Decision Science.

While implementing the action, analyze the data from the process or system change to see whether we are any further along in achieving our business objective. Assuming that the solution is not delivering, some tweaking or redesigning of the course of action is necessary, and another loop is launched. If this seems familiar, it is because *Agile* methods use this approach.

Resource-based strategy and decision science

Decision science and resource-based planning are intrinsically connected through their core objective: the optimal allocation of resources. This synergy is particularly evident in firms that adopt resource-based planning as their strategic backbone, making decision science an ideal complement to their planning and operational framework.

Resource-based planning is the school of strategic thought that states that you can't execute strategies if you don't have the tangible and intangible assets to make it happen. It posits that successful strategies require not only the possession of these assets but also their unique allocation in a manner that secures a competitive advantage. Decision science is concerned primarily with allocating resources to achieve specific outcomes, making it the ideal analytical engine for resource-based planning. Here's why.

Resource optimization

At the heart of resource-based planning is the principle of maximizing the utility of a firm's assets—whether physical, financial, human, or intangible assets like intellectual property. Decision science provides the methodologies and tools to analyze various decision-making scenarios, enabling firms to allocate their resources to maximize efficiency and effectiveness.

> *By applying decision science, firms can identify the most valuable use of their resources to achieve competitive advantage.*

Strategic alignment

Resource-based planning leverages a firm's unique resources and capabilities to carve out a competitive niche. Decision science supports this by offering a framework for making strategic decisions that align with the firm's core strengths and resources. Through techniques like decision analysis and optimization models, firms can evaluate the potential impact of different strategic options, ensuring that resources are directed toward initiatives that reinforce the firm's strategic position.

Risk management

Both resource-based planning and decision science recognize the uncertainty inherent in business environments. Decision science, emphasizing risk analysis, allows firms to quantify and manage risks associated with various resource allocation decisions. This risk-aware approach enables firms to make informed decisions that consider not only the expected returns but also the volatility and risk associated with different resource allocation strategies.

Enhanced decision-making

Decision science integrates qualitative and quantitative analysis to improve decision-making processes. For firms engaged in resource-based planning, this means being able to systematically evaluate the potential returns on investment for different resource allocation choices. The use of decision science models and tools can help firms navigate complex decisions by providing a structured approach to analyzing options and their implications. The quantitative methods to do so are presented in book 3.

Dynamic alignment

The competitive landscape and market conditions are ever-changing, requiring firms to dynamically adapt their strategies and resource allocations. Decision science equips firms with the analytical capabilities to respond to changes swiftly and effectively. By continuously assessing the value and contribution of resources in light of new information and market trends, firms can adjust their resource allocation decisions to maintain strategic agility.

In essence, decision science offers a robust framework and a set of analytical tools that enhance the resource-based planning process. By focusing on the optimal allocation of resources, decision science helps firms not only to maximize the value of their assets but also to sustain competitive advantage in a dynamic business environment. This alignment makes decision science particularly well-suited to firms prioritizing resource-based planning as a cornerstone of their strategic management approach. A new model for the organization is required to successfully combine these approaches.

The dynamic enterprise alignment model

When an organization wants to make decision science a core competency and have it underpin how decisions are made collaboratively, with data and expertise, a new model for envisioning an organization is required. Born out of research on enterprise relationship management, a framework for developing and managing the risk and performance of collaborative process networks can 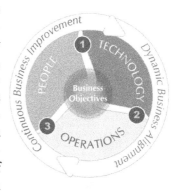 coordinate decision-making across organizational boundaries.

On the outside perimeter of the model is a continuous improvement/alignment cycle. Within the model are People, Operations, and Technology—representing the resources within a process or organization that can and should be leveraged to achieve results. The organization's vision and business objectives lie at the center of the model.

We measure he success of a resource-based strategy by its impact on the objective's performance. The DEA model encourages iterative reviews of resource allocation decisions on an evolutionary level through continuous business improvement yet also on a revolutionary level through dynamic business alignment, allowing an organization to pivot strategies to stay relevant (aligned) in its sector.

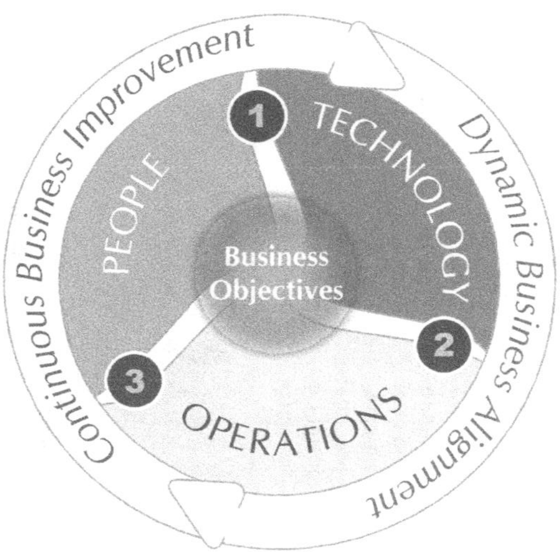

Technology includes all the processes, tools, applications and infrastructure necessary to acquire knowledge and support day-to-day operations and strategies.

Important considerations include:

- Do our people have skills
- Learning curves to get results
- Availability or lack of data
- Do we have the right software and systems

Business Objectives are goals that translate into a series of decisions and irrevocable allocations of resources to achieve desired outcomes

Operations include operational and administrative processes, performance management, quality management, physical technology used in the firm, a firm's plant and equipment, its geographic location, and its access to raw materials.

Important considerations include:

- Business Process Design
- Forecasting Accuracy
- Integration vs silos

Dynamic Business Alignment consists of the choices made by one firm or enterprise, that over time, help maintain the firm in sync with stakeholders and their environment. This has the objective of identifying and adopting business practices from indirect competitors or other industries to be able to innovate and make longer-term changes to remain a going concern.

People includes a firm's organizational structure, competency management, change management, employee development, formal reporting structure, formal and informal planning, controlling, and coordinating systems, as well as formal and informal relations among groups (either within the firm or between the firm and those in its environment.)

Important considerations include:

- Leveraging internal networks
- Resource and skills availability
- Cultural acceptability
- Compensation

Gaps (1, 2 & 3) are a lack of alignment between the objectives and processes. E.g. friction, slow response times, data not available / visible, conflict...

Continuous Business Improvement are regular and incremental improvement relating to organizational learning, day-to-day operational analysis, performance and quality assessments that improve alignment with stakeholder expectations (customers, management and the employees).

Figure 11: Optimizing resource allocation in the Dynamic Enterprise.

TABLE 3: CRITICAL SUCCESS FACTORS.

Model Component	Critical Success Factor	Org. Alignment	Strategic Alignment	Operational Alignment
Business Objective(S)	Make sure everybody agrees on the decision to be made.	Set a shared vision by involving all decision stakeholders.	Set business objectives to carry out shared vision.	Set operating and planning objectives to realize strategic objectives.
People	People within the organization are skilled and regularly communicate (laterally and vertically) about decisions and outcomes.	Create a culture and clear expectations that foster performance.	Ensure the skills, competencies, and internal networks are geared toward achieving long-term objectives	Foster a culture of rapid adoption of change and innovation in systems and processes.
Operations	Availability of resources, processes, and activities that seek to develop or leverage a firm's operational expertise in attaining business objectives.	Create an understanding of processes, capabilities, operating culture, principles, and assumptions.	Acquire competencies and develop tangible and intangible assets.	Develop performance, flexibility, and efficiency of business activities.

Model Component	Critical Success Factor	Org. Align-ment	Strategic Align-ment	Oper-ational Alignment
Technology	Leverage informa-tion and knowledge systems as enablers for opera-tional ex-cellence.	Full under-standing of the firm's business technolo-gy capabil-ities.	Alignment between business strategy, IS/IT strat-egy, and stakehold-er needs.	Rapid deployment of new systems and Innovation.
Gaps	System-atically addressing issues that prevent resources from inter-acting in a way that delivers value.	There is a lack of under-standing of the re-lationships between people, opera-tions, and technol-ogy.	The ob-jectives in one area do not support those in the others. (e.g., Budget Allocation Process to strategic initiatives).	The pro-cesses and activities in one area impede the perfor-mance of processes and activi-ties in the other 2.
Dynamic Business Alignment	Dynamic Alignment between the shared vision/ob-jectives of the organi-zation and stakehold-er expec-tations.	Under-stand customer expecta-tions and market-place require-ments.	Generate knowl-edge that enables the firm to respond rapidly to changing customer require-ments and market-place con-ditions.	Put the right infor-mation in the hands of the right people at the right time.

Model Component	Critical Success Factor	Org. Alignment	Strategic Alignment	Operational Alignment
Continuous Business Improvement	Regularly review business practices to eliminate nonvalue-adding activities and adopt new business practices that incrementally improve quality or speed.	Continuous integration of organizational learning/knowledge into strategy and operationS	Identification of improvement opportunities to better meet customer expectations	Implementing and monitoring business changes to assess benefits and impacts on organizational, strategic, and operational alignment.

Managing alignment on three levels

Organizational alignment implies that the firm's stakeholders (management team, employees, partners, clients, etc.) share a universal understanding of the objectives and issues facing the firm. Organizational alignment also means that the different sub-organizations must harmonize their objectives for the overall success of the firm (more on this in the next section). In most cases, these cultures, metrics, and objectives are at odds with each other, especially difficult when they are all interpretations of the same global business objectives using different filters and mental models.

> *Without a shared understanding of the issues and opportunities, each sub-organization will dictate its own priorities for its people, operations, and technology resources.*

Strategic alignment is based on two fundamental assumptions:

- *Economic performance* is directly related to the ability of management to create a strategic fit between the position of an organization in the competitive product market and the design of an appropriate administrative structure to support its execution.

- *Strategic fit is dynamic.* The choices made by one firm or enterprise (if fundamentally strategic) will, over time, evoke immediate actions that necessitate subsequent responses. Thus, strategic alignment is not an event but a continuous adaptation and change process.

Operational alignment implies that the firm's intellectual and physical resources aim to achieve specific operating objectives, such as quality, inventory control, customer service, etc. Operational alignment is often referred to as management by objective. It includes all the planning, monitoring, leading, and control processes such as budgets, marketing, and operating plans to be clearly linked to corporate strategy.

Aligning resources and closing gaps

When seen under the lens of the DEA model, decision science can help bridge gaps between people, technology, and operations by providing a structured framework to make informed and data-driven decisions. As a decision superhero, you will eventually stumble on ways to make the business smarter and smoother by identifying areas of friction or organizational opacity. A decision superhero will act as an integrator of knowledge by involving the right stakeholders and helping identify potential courses of action to improve or correct the situation. The DEA prescribes three integration gaps that, if addressed, allow a firm to operate more effectively:

1. *People/Technology gaps* exist when technology, systems, and application decisions are not enabling your people to do more.

2. *Technology/Operations gaps* exist when your systems and technology decisions do not reflect the organization's processes, strategies, or target business objectives.

3. *Operations/People gaps* exist when your people's capabilities do not reflect strategy or process requirements.

People/technology gaps

People/technology gaps arise when there's a misalignment between the tools and systems a company implements and the human workforce using them. The technology may be underutilized, too complex, or unsupportive of the tasks employees must perform.

A decision superhero can help by assessing the effectiveness of technology through user feedback, productivity metrics, and task analysis to develop decision models that determine whether investments in new technology align with human resource capabilities and needs. If the analysis reveals that the technology is underutilized or hindering performance, decision scientists can support IT and the business in evaluating and quantifying the benefits of modifications that better suit the needs and capabilities of the employees.

Technology/operations gaps

This gap manifests when the technological infrastructure does not support or enhance an organization's operational processes, strategies, or objectives. It could be that the technology is outdated, incompatible with new processes, or not scalable to meet future strategic goals.

Decision science employs optimization and simulation models that can analyze the fit between current technology and operational processes. The resulting models seek to predict future performance under various scenarios and identify where technology upgrades or changes are needed to support operational efficiency.

Suppose an organization's strategic objective is to reduce product delivery times. Decision science could use simulation models to test different technological solutions, like implementing an advanced logistics software system versus building a new distribution center, and determine how each solution might impact delivery times. Based on the outcomes, the company can make a data-driven decision on which technology to invest in.

Operations/people gaps

Operations/people gaps are present when there's a disconnect between the capabilities of the workforce and the operational demands of the organization's strategy or processes. This might be due to inadequate training, lack of necessary skills, cultural differences among regions, or being unaware of all the activities in a process and how they interact.

Decision science can support organizational change management efforts by identifying skill deficiencies and training needs through capability analysis, workforce analytics, and performance metrics. It can also apply forecasting models to predict future skill requirements based on strategic direction.

Regarding the human element, another area where decision superheroes shine is identifying communication and information flow issues using social/organizational network analysis. By statistically analyzing how internal networks operate, decision scientists can quantitatively recommend specific training programs, process changes, organizational modifications, or activities that improve communication and foster more fruitful relationships between stakeholders and the people on the front line.

Of course, these are just a few examples of the things that throw sand in the gears that we call gaps. Addressing these systematically as they arise will improve a firm's performance over time. The decision superhero is there to support making informed decisions that balance complex interdependencies and data-driven insights that economically improve alignment and performance across the company's people, technology, and operations.

Stakeholder alignment

According to John Hagel, in his book on web services strategy, any firm can be split up into three sub-businesses or sub-organizations with their own unique cultures, decision-making patterns, and problem-solving methodologies:

- *The people business* (sales and customer services) to sell and match available solutions to customer problems.

- *The production business* to produce at the lowest price and highest quality.

- *The innovation business* to design new products and services to sustain existing customers, meet the needs of new customers, or facilitate entry into a new market.

When faced with the same problem, the classic case is that all three sub-organizations will propose different solutions, which may not work together when taken as a whole. These three sub-organizations illustrate the different objectives they set for each other and what metrics they use to measure their success. Alignment becomes even more complex when you factor in the whole culture issue. In most cases, these cultures, metrics, and objectives are at odds with each other, especially difficult when they are all interpretations of the same global business objectives using different filters and mental models (Senge, 1990).

Same problem, **3 different solutions!**

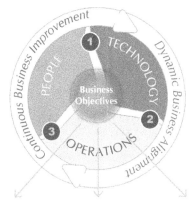

Mental Model	Customer Centric	Performance Centric	Innovation Centric
Skills	Direct Marketing	Operations	Product Innovation
Economics	Economies of scope	Economies of scale	Efficiencies of speed
Culture	Service-oriented	Cost conscious	Creative culture

Figure 12: Same problem, three different solutions!

In a sense, we need a level of abstraction. This usually happens by kicking solutions, making decisions, and modeling up one management level to assemble the parts into a coherent business strategy. Therefore, for alignment to occur, it is critical to consider the different cultures within an organization. Given that each internal constituency/organization has its own priorities and methodologies, understanding how these approaches diverge makes bridging the culture gaps feasible.

Therefore, working with stakeholders from all three sub-organizations becomes critical to develop a shared understanding of common issues. **It is equally important to note that without a shared understanding of the issues and opportunities, each sub-organization will dictate its own priorities for its People, Operations, and Technology resources.**

Business outcomes of the Dynamic Enterprise Approach

An organization that has successfully integrated a resource-based strategy in conjunction with its broader analytics strategy (decision science, business intelligence, data science) will have the following technical and managerial outcomes:

- People are motivated and share insights, have the skills to do the job, and have the backing of management.

- The technology and data are available to do analytics responsively.

- Operational decisions are supported by data and insight.

Figure 13: What success looks like after successful implementation of an analytics organization.

Silos versus integration

Silos represent a segregated structure where each department or unit operates independently, while a cohesive, process-oriented approach encourages collaboration across different departments.

Silos epitomize a traditional framework where departments function autonomously, with barriers to communication and collaboration. This segmentation often leads to a constrained exchange of information, with departments working towards individual objectives that may not align with the organization's collective goals.

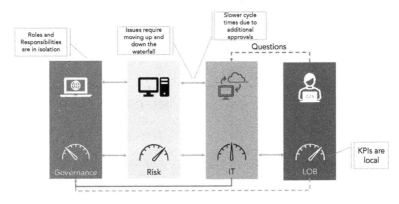

Figure 14: Silos and how to recognize them.

A process-oriented view fosters a culture of collaboration and inter-departmental synergy. Departments are no longer isolated entities but are part of a cohesive whole, working in concert toward shared objectives. This integrated model emphasizes streamlined processes that span multiple departments, encouraging the sharing of resources and a unified approach to achieving strategic goals.

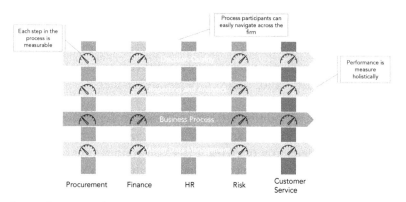

Figure 15: Integrated process view.

The transition from silos to horizontal integration represents a shift toward a more sophisticated and effective organizational paradigm.

TABLE 4: FROM SILO TO INTEGRATION.

SILO	→	INTEGRATION
Fragmented Systems	→	Integrated workflows and systems
Fragmented work: This is how WE do it in our group	→	Manage by process: This is how OUR organization does it
Slow and Error Prone Information	→	Readily available information + consistent data
Rigid Structures and Responsibilities (Not my job)	→	Accountability and Flexibility
Inward Focus on Performance	→	External Focus (Total Process Cycle)

Understanding the principles of silos and integration makes fitting data science fiction a much more effective prospect. Let's examine a few ways to organize around analytics and decision science to recognize our environment and perhaps plot an end-state.

Fitting decision science into the organization chart

There are many organizational structures in which a decision superhero can both exist and succeed, while in others, they can make the work much more challenging. As with all things, consider certain critical success factors need to when analyzing the organizational role of the decision scientist—both intended (that which we define on paper) and emerging (what actually ends up happening). The following should apply to a decision scientist:

- They have access to the decision-maker to be close to the decision.

- They are the person facilitating the process, or at least are a key influencer.

- They have adequate support and ensure management buy-in for the decision science (quantitative decision-making) approach.

If the decision-maker and the project team are organizationally well separated, it raises the question of where to locate the decision scientist. Robert Grossman and Kevin Siegel summarize the basic structures of decentralized, centralized, or a hybrid of these two, while Davenport and Harris suggest five similar structures. The goal is to better understand where people best fit in.

As an employee, you may not have a say as to where you fit into the organization chart, but at some point in your career, you will likely have an opportunity to influence the organization. To keep things simple, we organized these organizational structures into two broad categories: centralized and decentralized. Solid lines represent direct supervision and dashed lines support relationships.

Centralized organizational structures

If an organization's main objective is to have a consolidated and consistent process for generating insights, predictions, and forecasts using decision science, then it is important to consider a centralized approach. Centralized models come in different levels of implementation. A fully centralized organization *ensures that it follows a certain governance process and that the analytics process for generating results is auditable.* It can also entail having a centralized view of all the projects and initiatives in the same place, allowing for the efficient allocation of resources to the decisions and projects that require them the most. Of course, this requires a lot of mobilization from executive management and may be the product of a multi-year initiative.

On the other hand, perhaps a more gradual approach is to start with setting up a center of excellence that essentially consolidates best practices as well as identifies the key champions and practitioners. While this may be an easy way to get started, it does not enforce any of the best practices that it recommends. An intermediary solution to full centralization is to migrate from a center of excellence model to a consulting model, which allows the practitioners of the internal consulting group to both teach and apply methods consistently. Let's consider each one of these models separately.

A centralized model

The centralized approach groups all decision scientists in one place within the overall organization, as seen in the following figure:

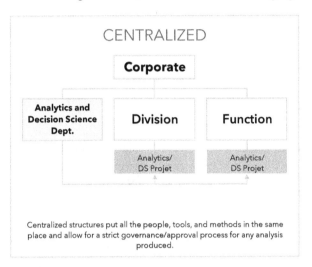

Figure 16: Centralized organizational structure.

Centralization brings together professionals who provide the same function for the organization. If they are located near the decision-maker, this can help communications for the decision itself. In general, the decision scientist is separated from the project, making it more difficult to facilitate work with the project team. Besides spatial separation, you may be in a different time zone or have different cultural values from the project team, which adds to challenges in communication.

Centralization facilitates quality control of analytic efforts. One of us worked for a consulting firm with the principle that when someone modified a model, they would get someone else to verify the modification was correct. Sometimes, when working alone, a consultant would have to bundle a fair amount of work before they sought verification. However, corrections can be made daily when working with a partner or a team. In a corporate setting, centralization allows for implementing a similar governance process that will ensure the quality of the analysis before making a decision.

This process incorporates a workflow that establishes a feedback loop between the stakeholders and the analytics/decision science team. This iterative communication ensures the analysis reaches a final form accepted by all parties and robust enough to inform decision-making.

The managerial advantage of the centralized structure is that it allows an organization's bad news to flow upward more directly and with less friction. Sometimes, centralization happens by default because the department consists of a sole practitioner. Needing more time to think, sole practitioners may respond, "*Let me go back to the office and kick this around with the staff to see what we should do.*" At least the staff meetings will be quick!

In all seriousness, a centralized approach allows for:

- Networking decision science practitioners, their knowledge, and best practices.

- Improve information flows

- Allow for introducing stage gates into the decision-making process

- Enforcing processes and best practices

A center of excellence model

As organizations increasingly recognize the value of decision science across various departments, it becomes clear that consolidating and sharing this expertise is crucial to avoid duplicating efforts. Establishing a Center of Excellence for decision science facilitates organizational-wide training on related subjects, standardizes data sources and methodologies, and evaluates decision-making performance, ensuring a more unified and efficient approach to decision science.

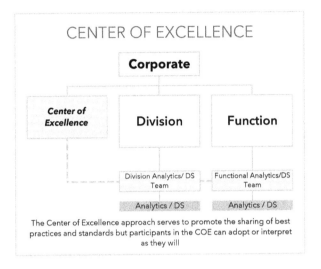

Figure 17: COE structure.

Organizing decision scientists in such a manner also sheds light on their requirements for professional growth and career advancement. A critical aspect of a Center of Excellence involves determining its funding model—specifically, whether it will charge for its services or if its expenses are overhead. This consideration is particularly vital in the early stages of adoption, as billing practices can influence its accessibility and use.

An internal consulting group model

If the number of personnel is large enough, a hybrid approach between the center of excellence and a fully centralized approach can be structured as an internal consulting group where decentralized individuals or teams can be available to work locally. Let's look at a quick representation of this:

Figure 18: Consulting structure.

A centralized team, benefiting from strong connections with local stakeholders, offers widespread support across the organization and acts as a readily available reservoir of resources, smoothing out fluctuations in demand. Often, this team is composed of more senior decision scientists, furnishing local teams with access to advanced support from seasoned experts. Additionally, this structure opens up a clear career trajectory for decision scientists within the organization by offering positions at different levels of seniority.

Decentralized organizational structures

Integrating analytics into decentralized organizational structures is valuable but is managed very locally. In other cases, it happens to be the natural starting point for the decision science journey. Let's take a look at a few of these structures.

A functional model

In a functional model, the decision science team is embedded into or reports to another function, such as strategy, finance, or operations, but is also identified as a function within the company.

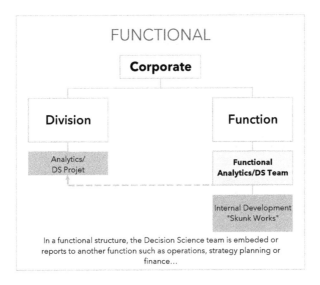

Figure 19: Functional structure.

This is the most classic starting point, and though this is decentralized, the DS group can be part of a governance process, just like finance or IT.

A fully decentralized model

In a fully decentralized model, every function and division has its own decision science expert, as we can see in the following figure:

Figure 20: Decentralized structure.

Being physically present offers numerous benefits, especially in an organization spread across multiple locations. As a decision-making expert, you play a crucial role in ensuring the process is adhered to correctly, which involves acting as a facilitator, elicitor, and coach. While it's possible to perform these roles remotely, being on-site tends to be more effective. Teams might operate on a part-time basis or not have a permanent status. For more complex projects, teams could become semi-permanent, existing until project completion. In such scenarios, having a decision scientist assigned to the project and stationed with the project team can be incredibly beneficial. This arrangement not only fosters strong relationships with team members, but also ensures that the decision scientist stays up-to-date with project progress.

A matrix organization model

A matrix-type organization, with functional units as well as business units, can improve the flow of information to the senior levels in very large and complex project-based organizations. Another aspect of a hybrid approach is to have decentralized teams at multiple levels within the organization.

Figure 21: Matrix structure.

This supports decision-making at the lower echelons, as well as decision science expertise to be brought to bear on recommendations as they work their way upward within the organization. This is common within the US Army for analytics roles. The work they do is reviewed by other analytics professionals serving at higher echelons of the organization, which supports DQ at the start of the process, as well as providing oversight from decision scientists and supporting reviews by more senior leaders.

What organization structure do you operate in?

Now that you have seen all the potential ways to organize decision science in a firm, consider which one best reflects where you work in its current state. This should give you a better sense of how to navigate management. The other takeaway is how you might want things to be like. With that in hand, knowing where to head can drive meaningful changes to process and structures.

Driving decision science adoption

Getting a firm to think differently is no easy task. After almost 25 years of tackling innovation and technology adoption projects, one thing is clear: everybody thinks it will be hard or impossible to get people to change or upskill. Sadly, people assume that groups are lazy or unwilling to adapt or assume the change is easy and obvious, but no one wants to do it. In part, there are good reasons for these apprehensions, usually rooted in a PTSD resulting from an underestimated and poorly executed change initiative that cost money, time, and delivered nothing.

Figure 22: The Lewin Change Model.

This seminal concept breaks-out change into before-transition-after states:

1. **Unfreeze**: This stage involves preparing the organization to accept that change is necessary. It's about breaking down the existing status quo before building a new way of operating. The goal is to create awareness of the need for change, which often requires dismantling psychological resistance and motivating stakeholders.

2. **Change (or Transition):** The transition phase can begin once the organization is unfrozen and open to change. This is the period where the actual transformation takes place. Here, organizations implement new processes, structures, or ways of working. It's a period of uncertainty and learning, and management needs to support and guide employees through this transition.

3. **Refreeze**: The final stage is where the new processes or organizational structures are solidified, and the organization is stable again. The new changes are institutionalized, and the organization develops new norms and policies. The goal is to ensure the changes are accepted and maintained over time.

In time and with the proper support from upper management, which includes regular investment in training and processes to onboard new people, the new way of doing things will become the way of doing things. It's like any change we make:

- **Offering rewards + aligning incentives** to reinforce the new state

- Identifying and **eliminating lingering barriers**

- Setting up **a feedback system** to collect and address any new issues or concerns

- **Modifying the organizational structure**, culture, and policies to align with the change and reinforce the new way of working

- Providing **employee training and support** to make employees comfortable and eliminate anxiety and doubts; and

- Establishing **quantitative metrics** to measure success and communicate progress.

If we take this idea and structure it into a series of strategies that build on each other to foster adoption, we get the adoption curve, the activities of which can be directly mapped to the Lewin model's three phases.

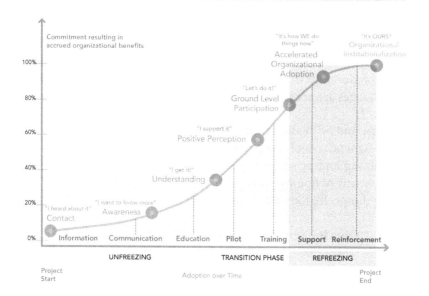

Figure 23: Mapping the Lewin change model to the adoption curve.

With a high-level understanding of the steps in the adoption curve and some of the important activities that support it, a decision superhero can develop strategies to get decision science accepted and adopted by the firm:

1. **Contact**: The people impacted by the project or new process receive information. Information is one-way and intended to prepare people that something is one the way.

2. **Awareness**: When people are aware something is coming, they will want to get more information. This is when a two-way conversation happens between the initiators of transformation and those impacted. This is an opportunity to update or adapt the base strategy.

3. **Understanding**: Education and executive primers on the fundamentals are provided, and individuals begin to understand what the change entails and why it is necessary.

4. **Positive Perception**: The best way to sell something is to see it used successfully. After a successful pilot/demonstration,

individuals start to see the potential benefits and gains from the change, developing a favorable view of the move.

5. **Ground-Level Participation**: Members of the organization may experiment with the change, selecting use cases to test its impact.

6. **Adoption**: Changes are rolled out and people progressively learn how to operate in their new roles and workflows. This is a learning and grace period where performance may be off, but it should be temporary if training and skill development resources are readily available.

7. **Institutionalization**: The change becomes part of the standard operating procedures and fully integrates into the organization.

8. **Internalization/Commitment**: Finally, the change is fully embraced and becomes a part of the organizational culture; individuals are committed to maintaining the change.

The actual transition starts at the understanding step and follows on to the ground-level participation step. Transition is very specific, meaning a change from one way of doing things to another. This does not necessarily mean that it is not messy. It often is.

Getting started on the right foot is very important if you want the adoption of decision science, analytics, or data science to succeed. By this, we mean addressing the gap in understanding between how the business and the technology group frame the problem and the benefit. It's a classic refrain to hear the business complain that IT is rigid and uncooperative and IT complain that the business is unrealistic in its expectations or demands. Figure 22 shows that the gap is the greatest at the beginning due to a higher degree of uncertainty about what the other party might expect.

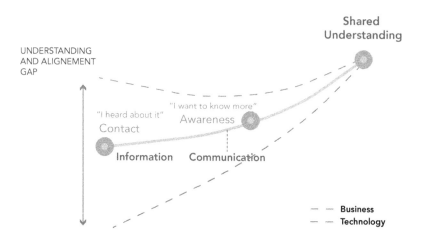

Figure 24: Managing the gap between business and technology.

With cyber-security being what it is, the technology group will have requirements and want to understand what you are trying to accomplish. Objectively, a decision superhero should work with an IT business analyst to understand the impacts of what he is proposing regarding technology or process modifications. The technology group may ask the decision superhero to do the *voodoo, that he do,* and produce a financial or economic case, which can become a sales tool for the project.

The aim of presenting these simple yet fundamental concepts in business transformation is to guide decision superheroes, leaders, and change managers on successfully implementing change by supporting their teams through these stages.

Effective use of such a model entails communicating clearly about the changes, providing support and resources for individuals to adapt, and fostering a positive environment where feedback is encouraged and contributions are recognized, all with the goal of moving individuals along the commitment curve from awareness to full commitment.

Key takeaways

- In this chapter, we focused on the ideas and frameworks that would allow a firm to make quantitative decisions easier and the insight used to make them widely available to the people who need them.

- Are decision superheroes bought or made? Machiavelli always made the point that your most effective lieutenants are the ones you foster internally. However, if the need is immediate, resorting to consultants and knowledge transfer/training can address immediate issues.

- An organization's strategic framework must align with the idea of optimal resource allocation for decision science to be a significant competitive advantage.

- Decision science provides a quantitative and intellectual framework for closing performance and alignment gaps within the Dynamic Enterprise Model.

- How to differentiate situations when decisions and projects are happening in silos rather than in an integrated way that has all the important stakeholders contributing.

- Having the right organizational structure matters, but so do internal networks that share knowledge and insights on how some problems have successfully (or unsuccessfully) tackled in the past.

- The Lewin change model and the adoption curve provide a framework for deploying new decision-making processes and their supporting analysis activities in a positively perceived and economically effective way.

The next chapter will focus on how decision superheroes can effectively collaborate with their equally important data superhero counterparts.

FOUR

When Data and Decision Superheroes Meet

Sometimes, a decision superhero cannot tackle a problem alone and needs to bring in some *data superheroes* to get the job done. Data superheroes, often known in the real world as data scientists, use their magical powers to cleanse and make sense of insane amounts of data to find insights. When decision superheroes can access cleaner and more detailed analysis from their data colleagues, they can build powerful predictive models that help decision-makers better understand what awaits.

In the previous chapter, we focused on better understanding the decision scientist. In this chapter, we will extend the same courtesy to our colleagues in data science to catalyze a conversation on how these roles are similar and different. We will briefly explore how to combine both disciplines to create more value for a firm, and how data science can better support the decision science function within a firm.

Decision science is to little data what data science is to big data

Much has been written over the years about how organizations are sitting on a wealth of data, even going so far as calling data the "new oil." Data is important, but what do you do when you don't have it? Decision science addresses this by compensating for missing data by working with experts and probabilistic modeling. Therefore, it is useful to differentiate between little and big data applications to know where to find complementarity when assembling a project team.

In Figure 25, we outline the knowledge areas that are both distinct and complimentary for Data and Decision Superheroes.

Figure 25: Decision versus data science skill sets.

Any business professional who understands probability and how to model it will have a measurable impact on the organization's performance.

The models are the same but what is expected is different

When talking to a data scientist and a decision scientist, you will realize that, though the basic definition of a model is the same, its applicability and expectations of the model are different. A model is simply a mathematical or visual representation of a system, problem, or idea. Every model has inputs, an algorithm, and outputs. In this regard, models are the same. Where they differ is in their origins and their objective functions. Data science models tend to be used to produce judgments while decision science models, to quote our good friend Rob Brown, are *"the irrevocable allocation of resources to achieve an intended goal."* The following table contrasts data and decision model applications. The main differences are how the models are derived, their relative opportunity cost, and the scale/frequency of the decisions.

TABLE 5: OPERATIONAL INNOVATION VERSUS STRATEGIC INNOVATION.

Sector	Data Science Applications	Decision Science Applications
Banking / Insurance	Credit scoring	Revenue forecasting
	Anomaly detection	Scenario planning
	Loan analytics (Prob. Of Default and Loss given Default Calculations)	Setting risk tolerances
		Process improvement
	Process automation	
	Micro segmentation/ targeting	
Extractives (Oil And Gas, Mining, Etc.)	Spectral analysis of aerial pictures to identify reserves	Reserves forecasting
		Production process optimization
	Optimize process operating parameters (for example, minimize the risk of a blowout)	Business development
		Project risk analysis
	Predictive maintenance	Mergers and acquisitions
Finance	Algorithmic trading	Fundamental analysis
	Real-time asset valuations	Business valuations
	Real-time pricing	
	Automatic portfolio rebalancing	

Sector	Data Science Applications	Decision Science Applications
Content Delivery (Youtube, Google, Tiktok, Netflix, Etc.)	Image recognition Image auto tagging Ad targeting Recommendation engines Natural Language Processing (auto captions)	System capacity planning Cyber risk quantification Mergers and acquisitions
Retail And E-Commerce	Pattern analysis (for example, shopping basket analysis) Product-level demand forecasts Optimize promotions Product recommendation engines Website experience optimization	Forecasting revenue and expenses Project portfolio optimization Marketing analytics Process improvement Comparing strategic alternatives
Utilities	Predictive maintenance Optimize process operating parameters	CAPEX decisions Project risk analysis Business development

The origins of a model for a data scientist are the inferences they can extract from a data set using algorithms. A model that produces a judgment can be derived using these inferences. For example, interesting relationships such as diapers selling beer, recommendation engines promoting products that may appeal to you, and so on. So, we go from the data model in this perspective. This is a very useful approach when you have a constant stream of data to throw at the problem, as with many large organizations, especially financial services, such as credit card and insurance companies and banks.

One of the main reasons data-driven models are so attractive is their ability to efficiently treat high volumes of important but extremely repetitive analyses that computers can produce more consistently. This makes perfect sense because it's cheaper, faster, and more reliable than the human equivalent. These are very important and help to improve an organization's profitability and risk profile. When put into production within the confines of a business process, data-driven models make decisions using classifications derived through these inferences. For example, approve/decline, spam/not spam, fraud/no fraud.

Furthermore, due to the sheer volume, automation of repetitive business decisions has a low opportunity cost associated with each decision. Unfortunately, if a model becomes deprecated (stops making good predictions), then you can negate any gains made up until then in one fell swoop or a slow decline. Always remember that when the market goes bananas, trading systems are shut down. Sometimes it's the market and sometimes it's because deprecated models are going to war with algorithmic traders.

Decision science takes a different approach by focusing on achieving objectives with limited resources rather than leveraging insights found in data. The question must stem from a need because no direct data exists as you would find in a machine learning model. This is the other foundational difference between data and decision sciences. Thus, the starting point is a statement of a desirable outcome. In basic terms, we have options and we can't have all of them—we must make choices. If I do A, I will let go of options B and C. Because of the relatively high utility and the decision being almost exclusively an opportunity cost problem, the relative weighing of each option being able to deliver the most value is the name of the game. In addition to your dusty economic books that you may or may not have read in school, the concept of opportunity cost can be found everywhere in the real world, such as picking the right mate, what to have for lunch, which car to buy, build or buy, move or stay.

Though you can have decisions that happen frequently in decision science, they will never be on the scale of a transactional system in a bank. However, they will have the advantage of being relatively simple and low impact, such as knowing what to have for lunch.

> As a general rule, decision frequency is inversely related to its cost, which simply translates into choose wisely, my friend.

One of the notable differences between these disciplines is the approach to solving problems. Data science is a grassroots data approach while decision science is more top-down. We can also explain this through the scientific reasoning approach.

What if I want to know more about Machine Learning and Data Science?

Some of you may want to know more about machine learning ideas, tools, and practices to better understand the capabilities of the data superhero, and there are two awesome resources. Alex Gutman and Jordan Goldmeier's book "Becoming a Data Head" is a broad overview of the methods and tools of machine learning, a perfect primer for anyone seeking to learn the ML lingo. Another great book is "Julia for Machine Learning" by Zacharias Voulgaris, which covers dimensionality reduction (knowing what variables to focus on), visualization, and both supervised and unsupervised machine learning.

When data science can be built out into a decision model

Generally, data science is concerned with looking at piles of data and identifying relationships (be it correlations or causal) and/or inferring via machine learning a model from the data using algorithms to make policy improvements. This inductive approach focuses on investigating—or, if you prefer, mining for little nuggets of gold.

Decision science, on the other hand, approaches the problem differently by starting with a question or a problem to be solved. This is known as a *deductive approach* and will guide what needs to be understood to construct an effective decision-making model. The starting point for a decision scientist is getting the model right while a data scientist is concerned with getting the data right first and figuring out the model after.

The good news is that though these two approaches may seem at odds with each other, they support and feed off each other in fairly virtuous ways. From a business perspective, decision science will often dictate or at least identify where data science efforts can yield the highest return on information. Data science is very good at opportunity identification. It can look at data patterns and unearth opportunities and relationships that have defied analysts' eyes, but what to do with that information? That's when the ball gets thrown back to the decision scientists to work with the business to identify and select the alternative with the best likelihood of success.

The data scientist is the one with the lantern and the decision scientist is the one with the map. In a dark maze, you need both.

Inductive versus deductive approaches to solving problems

Data science looks at reams of structured and unstructured data to develop theories on the mechanics of the underlying process or system generating the data. This is an inductive approach because the work serves to develop a theory on how things work but cannot easily answer the question *Why?* This is not to say that you cannot be inductive and prove theories by using A/B testing (or some other validation method), but generally, the work at this level is exploratory in nature and the goal is to dig up insights into the data.

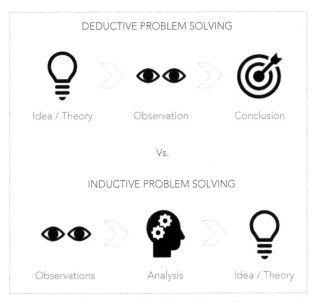

Figure 26: Contrasting inductive and deductive problem-solving approaches.

On the other hand, decision science starts out deductive by postulating an idea or a set of alternatives (**courses of action**) and running a simulation to see whether our understanding of the system is correct. Triangulation of mathematical methods can give a reasonable range in which your model should operate and thus allow you to study its behavior with peace of mind.

Figure 27: Combining the data and decision science problem approaches.

What is remarkable is that data science and decision science are two sides of the same equation. By understanding this unique relationship between the disciplines, we quickly see how they can reinforce each other. It's almost a dialectic approach, switching back and forth between both reasoning paradigms. *From the general to the specific and back again.*

For example, a decision model's relationships can be refined or include drivers, once unthought of, that were surfaced using a machine learning algorithm. Conversely, a decision science model can bootstrap a machine learning model when insufficient data is available. Furthermore, ML can be used for a more in-depth analysis of the model's simulation results, pointing out clusters and collisions deserving further study by the subject matter expert or decision-maker.

All this leads to the interesting question of how these two disciplines can interact at the business level for the greater benefit of all.

Decision and data superheroes unite!

Data and decision science form an interesting ecosystem of complementary ideas and techniques supporting each other in achieving both global- and area-specific objectives. Both the data and decision science models for solving problems have strategic and operational applications, which in turn are also interrelated.

At the strategic level, the name of the game is competitive advantage. This can take the form of knowing something others don't, being quicker, or being cheaper. Either discipline can bolster these generic strategies. Data science allows decision-makers to ponder patterns in data and see whether there may be business opportunities therein, while decision science allows you to look at these opportunities amongst each other and select the best one within the context of what is going on in the business. This involves collecting and analyzing large datasets to uncover trends, patterns, and insights that can inform strategic decisions. Techniques used include predictive analytics, clustering, and regression analysis.

Decision science in the strategic domain involves generating various strategic ideas or scenarios, evaluating their potential outcomes, and selecting the best course of action. This is done using techniques such as scenario analysis, decision trees, and cost-benefit analysis. Decision science can considerably augment its credibility and time to answer by engaging a data scientist for the data component of your modeling project.

> *Combining decision science and data science help organizations gain a competitive advantage.*

Data science provides the analytical insights needed to identify opportunities, while decision science evaluates these opportunities to determine the best strategic actions. Together, they enable organizations to make informed, data-driven strategic decisions that differentiate them from competitors.

Data and Decision Sciences, a case for marketing effectiveness!

While working for an auto and motorcycle financing firm in Asia, we were tasked with figuring out strategies to improve the profitability of the book of loans. Using exploratory analysis, we kicked off the process by identifying characteristics and features of borrowers that predicted whether or not a loan would default. To do so, we classified each past loan as either complete or defaulted (this classification is necessary for supervised machine learning) and were able to identify key relationships and predictors that could be used to build a forecast and drive modifications to the loan approval process. In this project, we used a machine learning/data-driven approach to explore the relationships and a decision science approach that used those insights to build a simulated forecasting model that measured how implementing the policy changes would impact business performance over time.

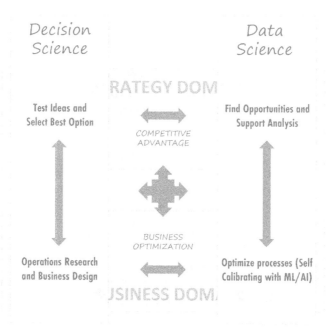

Figure 28: Systemizing data and decision science into a process.

In the business domain, decision science uses many classic mathematical techniques such as operations research, process modeling, simulation, and mathematical modeling to find better and more optimal ways of doing things while data superheroes will perform automation, machine learning, and potentially, optimization to get a predefined data-driven process working better. We can easily imagine one group of superheroes designing a process and another automating it and putting it on an optimization auto-pilot!

Why does a decision superhero need a data superhero?

It goes without saying that being a decision superhero means you get data, but sometimes, there is so much that you *actually* need an expert to make sense of what you are looking at. This is where a data superhero can provide invaluable data wrangling and exploratory analysis expertise that can translate into considerable time, effort, and scope savings.

Data wrangling

A data superhero must be skilled at data wrangling. Wrangling of data is mapping from the original data into a new data model or structure required for analysis or further use. The concept of the idea of bringing order from data chaos appears in the following figure:

Figure 29. Data Wrangling (by Christian Jensen, with permission).

A simple example would be parsing name data from a single text string into last name and first name fields. Typical school textbook homework problems provide straightforward data to allow problem solutions. This is not how we encounter data in practice. The data model is often different from what is needed, and the data often contains entries that are in error. Most data scientists spend more than 80% of their time getting data ready for use. This is sometimes called **data munging** or **data wrangling**. Merriam-Webster says munging is perhaps an alteration of the obsolete mange to eat, from Middle English *mangen*, from Middle French *mangier*.

Wrangling can only happen if you have data to wrangle. This is written from the perspective that the data exists in your organization and there are no legal barriers to its use. In some cases, you may need to pay vendors to use their data, and furthermore, some data may be used only in certain ways. For example, restrictions vary under different levels of government but still dictate how personal data about employees may be used or shared.

Data wrangling consists of several steps. Even if every analyst on the planet should be comfortable with notions of wrangling data, as mentioned earlier, there comes a time when you need an expert to get the ball rolling and help with some of the more technical aspects of the data problem.

- **Discovering**: This is understanding the data. Often data was collected for reasons other than the new use. This includes understanding the definition and intent of each field. For example, you may receive data on the availability of a particular facility. It likely is important to know whether the uptime data includes planned maintenance periods or only reflects availability during normal operating times.

- **Structuring**: Data may have been captured using a different data model and you need to map the extant fields to the new ones you will need. You may also have unstructured data that has not been fit to any data model. Often, you will parse this data to look for the information you need. For example, you want to look for part numbers. This structuring is normally done in conjunction with data extraction, the process of retrieving the data from where it was stored.

- **Cleaning**: The data will almost certainly have inaccuracies. There may be duplicate entries, partial entries, and spurious entries. If your data was unstructured initially, the structuring step may have been less successful than you would like. Besides these data errors, you may want to remove valid data that is not applicable. If you pulled data from your company's North American sales database, but you are only interested in Canadian sales, you will want to remove the Mexican and US sales data even though the data are valid for other purposes.

- **Enriching**: The idea of enriching is improving the information value of the data by combining it with other data. This might be simply adding additional data to improve accuracy, or it might be for a specific need. Canadian sales data might have gaps when the system failed to record the information in certain periods. However, the information might be recoverable from tax records and shipping records.

- **Validating**: While you have cleaned your data, that does not guarantee it is all valid. You might have a customer who routinely buys between 10 and 20 units but, in one period, instead buys 9,999. That looks invalid. And if your Canadian data is retail gasoline sales in liters, be sure you are aware of that if you were thinking of US gasoline sales in gallons. Validation is generally done by scripts applying rules.

- **Publishing**: This makes the data available for further use by a person or process. It should also include documentation of what was done in data wrangling and explain the data model.

- **Curation**: The idea of curating the data is to ensure the published data are usefully preserved over the long term. Curation will not be necessary for a single study.

These data skills distinguish a decision superhero's need for a data superhero. A decision scientist may need to wrangle data, but not often with the same degree of challenge or purpose. A data scientist uses stats, machine learning, and math for pattern identification from data and deep expertise on a specific problem scope, while decision scientists operate at a more holistic level and use data to inform decisions with data-based mathematical predictions. These professional areas still need to frame the issue, manage the project, and present the results. The presentation may differ in that the data scientist is more likely to provide ongoing analysis and is more likely to be dealing with larger data sets and using more detailed and scientific visualizations, while a decision scientist has a broader internal audience and will need to present results in terms of business outcomes using more conventional business charts and dashboards.

Exploratory analysis/opportunity identification

Because data superheroes are skilled at finding models and hidden relationships in data sets using inductive methods, they are particularly adept at helping decision scientists create models. Exploratory methods include using a suite of statistical measures and visualizations to identify patterns and formulate hypotheses that can be explored for value by the business. Classic examples of exploratory analysis of historical data include visualizing trends over time with visualizations, calculating correlation, aggregating and pivoting, regression analysis, scatter plots, and such. Languages such as Julia, Python, Matlab, or R can be quite effective for the command line avenger, or GRETL, Tableau, PowerBI, or TIBCO Spotfire for those who want to employ a visual approach to find patterns, relationships, and trends.

Using exploratory methods, data scientists can find scraps of interesting information that can lead to formulating business strategies to capitalize on the found insight. Exploratory analysis begets more exploratory analysis, so it is important to have an idea of what to look for, which should begin as a question from the decision scientist.

When no normative model exists, then one must be derived. This is where data-driven methods can help inform the decision model development process.

Why does a data superhero need a decision superhero?

A data superhero can also benefit from having a decision superhero in their corner. Decision superheroes are very good at determining where you get the most value for time. They tend to be close to the business and know the current strategic and technological orientation of the firm. Because analytics and data science investments can take time and the innate desire to boil the ocean, funding and interesting projects can become scarce due to a lack of immediate results.

Here are a few reasons why a decision superhero can support the data science function to avoid such roadblocks.

Targeting data science for more bang for your buck

Too many good ideas are just as bad as not enough. This is the same thing that happens when we go to a restaurant with a 30-page menu of a bunch of things we would like to eat. In this situation, the time to eat, amount to eat, and budget constrain us. So, we need to figure out what will give us the best experience, all while accounting for these constraints with a ridiculous amount of options to consider. For anybody who has ever experienced this, this can be quite excruciating. A decision scientist is like that helpful friend or waiter who points you to the items on the menu that are worth trying. This allows us to focus on the best opportunities, which we mean are the ones with the highest likelihood of delivering superior results within time, budget, and preference constraints.

Political cover

At the annual *2019 Decision Analysis Affinity Group (DAAG)* conference, a speaker from a well-known consulting company told a cautionary tale about how his firm lost a deal by selling data science instead of a solution targeting an actual need. Generally, a situation to avoid.

The technical members of the team were data scientists, and they focused on the benefits they could deliver on improved processing speed on the client's massive data. The potential client's representatives had asked at several different points how the improved speed would provide value to their business. The consulting pitch team focused on the data and technical details but did not address the question of why this was a good idea. Sure, increased speed would be better, but it was not free, and was it worth the cost? The client was faced with a decision and the data scientists had not understood their perspective.

Understanding that advanced analytics is not a monolith implies that it breaks down into several domains:

- Business Problem (Question) Framing
- Analytics Problem Framing
- Data
- Methodology (Approach) Selection
- Model Building
- Deployment
- Model Life Cycle Management

A quick analysis suggests that the consulting team skipped the first domain. They weren't listening and failed to understand the business' perspective, which is a cardinal sin in consulting. They may have gotten the second domain correct but without correctly justifying the cost or tying it back to a stated business need. Alternatively, since they weren't paying attention, they may have gotten the second domain incorrect and been working on the wrong problem all along. Had a decision science practitioner been involved who bridged the gap between the need and the options to solve the problem, the project might have gone forward. Unfortunately, in the end, it was a waste of time for all involved.

Key takeaways

- Once you understand what is needed and how to approach it, data science and decision science may take different paths. Data requirements and sources can be identified when the analysis concept is understood. (These domains are applied iteratively. Examining available data and finding it lacking may cause you to replan the analytics approach.) The other domain tasks will differ if you have big data or little data.

- If you are working with big data, you might decide on a machine learning approach—train and evaluate the model, deploy it, and then have a plan to periodically reassess its performance. If you are working on little data, you might build a simulation, test it, and then deploy and manage it over time. Both workstreams agree with the CAP domains, but the nature of the tools is quite different.

- Besides this either/or approach, the two paths can be used in a complementary fashion. If you have big data and decide to build an artificial neural net, you can also take a sample of the data. The sample now has little data, and other techniques can be used.

- This little data modeling effort offers several potential benefits. The results will be available quickly if it is a simple statistical model. If the big data path has problems, the little data work will provide an answer, although perhaps one that is less precise. The two paths provide dissimilar approaches, so if they produce differing results, that is a valuable warning. The reason for it should be determined. If the results are similar, they validate both work paths and raise your confidence.

- Data science and decision science share some skills and capabilities but differ in others. Data science and decision science complement each other. Using the right approaches for your circumstances is the key. We will take that up in the next chapter.

- Conversely, the decision science group may commission studies and obtain detailed data from their data science counterparts to build a model that helps leadership make decisions such as how to allocate resources, set priorities, right-size budgets, and make forecasts. If the data and decision science functions can collaborate successfully, both functions benefit. Data science benefits from having strategic objectives for their work, while the decision science groups benefit from having better and more accurate data for their decision models.

Over the last couple of pages, we have taken a deeper look at the role of the data scientist and how they can support the decision science function within an organization to improve the quality of predictions and decisions leadership makes. In the next chapter, we will explore some ways we can frame this process of working together and some methodological tools for framing the problem correctly upfront so that we can start building our models with the right understanding in mind.

FIVE

Decisions are Models and Models are Decisions

The foundational toolset for every decision scientist and analyst is modeling. That is, translating an idea or a system into math you can play with. By clearly understanding how decision models are generated at both the intellectual and practical levels, the analyst can solve problems and get solutions (models) implemented in R, Python, Julia, and Excel for maximum impact.

In the following pages, we will explore the true nature of models and how to think about them. Everything we do in decision-making is based on some sort of model. Just like recognizing data and programming patterns, it is a true superpower to think in terms of models. Models can represent both concrete and abstract things. From airplanes to profitability, a model can bring an idea into a form that is shareable and consumable by many.

Though the notion of models has been around since ancient times, techniques for building models and expressing results coherently have evolved. Now we can use simulation, machine learning, as well as a host of other tools to bring models to life and understand the hidden dynamics of the problem under study. The secret is knowing when and where each approach and model type will have the most impact and generate the best result economically, both in terms of money and time.

By the end of this chapter, you will have a high-level understanding of models and how to go about scoping their purpose and managing the results.

Universal model constructors

Core idea: *Models are visual or mathematical abstractions of reality used to explain and analyze a problem or phenomenon.* Every model has three constructors or it's not a model:

- **Inputs**: Ranges or single input values

- **Model or Algorithm**: Set of mathematical relationships, f(x), e.g., *Profit = Revenue - Expenses*

- **Outputs**: The variable(s) of interest you want to analyze: *Profit and Profit Margin*

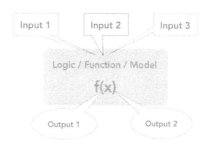

Figure 30: Components of a model.

Every model must/can do the following:

- Have a clearly defined question or purpose to answer.

- Vary conditions and assumptions to test scenarios.

- Have at least one input and one output. Most models have many inputs and many outputs. There is no formal relationship beyond having at least one of each.

- Reflect reality as closely as economically possible. The logic and formulae need to be valid. *Validation of this is a big driver of successful models.*

- Account for user inputs (decision variables under the modeler's control) along with uncertain ones, which are represented as distributions.

- Produce meaningful outputs and solve the question the model was designed to answer.

Models exist in different forms and places

Models make our understanding of the world possible by giving us the tools to simplify it in a way that makes sense in our heads. Of course, you could point out that all models started as mental models, but physical models are observable in the real world. In Figure 29, as you move from left to right, the model types transition from their most abstract form to their most concrete and defined form. The problem with this type of classification is that it can muddy the waters in a conversation and is best suited to you, the decision superhero, to use as a framework for eliciting and searching out solution paths. *Just be careful with classifying models because smart people tend to get caught up in the definitions.*

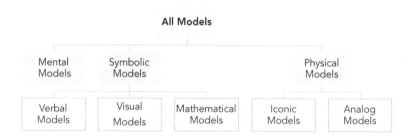

Figure 31: Model Types and Classifications.[7]

Because the starting point for all models is the mental model, let that be our starting point for the conversation.

Mental models define the world we see

When we think about reality, what are we really doing? As we step into solving any problem involving the real world, we must find a way to describe, explain, or even predict the part of reality in which the problem is embedded. The one fact that stands out right at the beginning is that our thinking goes on inside our head and we do not have the capacity to get irrelevant chunks of reality into our head. And since reality always remains external, the only thing we can do is get thoughts about reality into our brains. Our thoughts, then, are abstractions from reality.[8]

At some point, models need to interact with the real world. Given that a model is a simplified but useful approximation of some reality or idea that stems from a mix of experiences, education, culture, and sensory perceptions (taste, smell, touch, sound, and sight), it is sometimes necessary to put it into a form (in our case mathematical) that is understandable by others.

[7] Adapted from Springer, Herlihy, Mall, & Beggs, Statistical Inference: Volume 2 of Mathematics for Management Series, 1968

[8] (Springer, Herlihy, Mall, & Beggs, Statistical Inference : Volume 2 of Mathematics for Management Series, 1968).

The modeling process entails iterating and adapting the model as new information becomes available, leading to either action or concluding that the model, ideas, or assumptions are invalid, and therefore we abandon the decision.

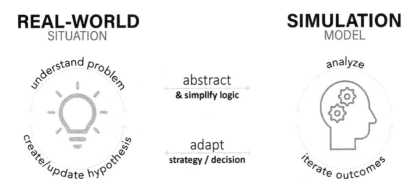

REAL-WORLD
SITUATION

SIMULATION
MODEL

understand problem

create/update hypothesis

abstract
& simplify logic

adapt
strategy / decision

analyze

iterate outcomes

Figure 32: Relationship between models, simulation and the real world.

The simplest example of an abstract mathematical model that we can all relate to, which exists in our head and not in the "real" world, is **PROFIT = REVENUE - EXPENSES.**

TABLE 6: EXAMPLE OF A SIMPLE MODEL AND ITS COMPONENTS.

Model Constructors	Objects/Components and Relationships
Inputs	Total Revenue and Total Expenses
Model (Formula or Algorithm)	Revenue - Expenses
Outputs (Objective Function)	Profit

Now, most of you are probably thinking, "There is nothing abstract about profit!!!" I get it, but you just need to apply the right yardstick. To demonstrate that profit is an abstract concept, consider that even though you know *exactly* what profit means, you will never find a pile of profit in nature, nor can you step in it on a warm spring day in the park. *You can only see it in your mind by **differencing** two numbers.*

Symbolic models

Imagine a complex problem that's hard to picture with physical objects or direct comparisons. To tackle this, we turn to symbolic or mathematical models. These models use symbols, such as Greek letters, parts of the regular alphabet, or simple names, to represent the problem's different parts. It's like translating a complicated situation into a language of mathematics that we can manipulate to find solutions.

For instance, while a small-scale model airplane (an iconic model) helps us understand the design of an aircraft, and a fuel gauge (an analog model) gives us an idea of how much fuel is left without showing us the actual fuel tank, symbolic models are different. They come into play for scenarios that are too intricate or abstract for physical models. We use equations and symbols to represent different factors at play, like stock prices or temperature changes, to understand the behavior of the stock market or predict weather patterns, which we can then analyze to make predictions.

Now, when we talk about verbal models, we're referring to the use of language to describe or record information. This could be anything from a written report, an email conversation, website content, or data stored in a database. Text mining, a technique in analytics, is all about exploring these verbal models to uncover valuable insights. *However, when we process this information, we often form our own mental picture or model of the situation described by these verbal models, sometimes without even realizing it.*

It's crucial to distinguish between these mental models and the actual, external information or models. Philosophers have a term for when we confuse our internal understanding or perception of something with the actual reality: they call it hypostatization. This concept highlights the importance of recognizing the difference between what's happening in our minds and what's happening in the world outside[9].

Other symbolic models include visual models and mathematical models. Mathematical models consist of a set of one or more mathematical equations. Graphs and diagrams are models. They aid in visualizing the system and its performance. In some cases, they are the "same," meaning that visual symbols and their mathematical relationships are both modeled using the same tool. Symbolic visual models can take many forms, including workflows, flow charts, decision trees, plant layouts, Gantt charts, and the venerable influence diagram.

Iconic and analog models

In the realm of modeling, the most tangible category encompasses both Iconic and Analog models. Iconic models serve as physical or conceptual representations of specific items, like cars, houses, or even plants. They can come to life in two dimensions (2D) - think photos, drawings, or blueprints - or in three dimensions (3D), such as a miniature airplane or a detailed architectural model. Imagine a paper airplane; it intriguingly straddles the line between 2D and 3D, starting as a flat piece of paper but ending up as a three-dimensional object you can hold.

On the other hand, Analog models operate a bit differently. They're not about representing objects but rather about illustrating processes or ideas in a form we can easily visualize and understand. For example, a demand curve in economics shows the relationship between

---

[9] (Springer, Herlihy, Mall, & Beggs, Statistical Inference : Volume 2 of Mathematics for Management Series, 1968).

a product's price and its demand, or a histogram that displays data distribution. These models translate complex processes into visual diagrams that clarify the underlying principles. Tools like Microsoft Visio are invaluable for creating these visual representations. Visio stands out due to its comprehensive set of features that cater to a wide variety of visual modeling needs, from simple diagrams to complex process maps, making it a top choice for those needing to craft detailed analog models.

Models represent (and are) systems as well

Think of a model as a way to represent something complex, like a system or a part of that system, so we can understand or study it better. These models can be physical, like miniature versions of things we can see and touch, and they're used all over the place for different purposes.

For instance, in wind tunnel tests, small-scale *models* of airplanes or cars are used to study how wind moves around them. Pilots train with flight simulators that mimic the experience of flying a real plane, helping them prepare for actual flights without leaving the ground. These physical models can represent a wide array of systems, including transportation, production, supply chains, and even health or education systems.

Now, when we talk about a system in this context, we're referring to *a group of objects or components that work together towards a common goal*. These systems can be just ideas in someone's head (abstract) or real, tangible things. The components of a system are the individual parts or objects that make up the system, and they usually need to be managed or controlled in some way. This control can be external, like a person making decisions, or it can happen automatically within the system itself, like an airplane's autopilot.

Systems operate within an environment that affects how they perform. This environment includes factors we can control and some that we can't. We use simulation and optimization to ensure these systems work as well as possible, especially when we can't predict everything that might happen. These methods allow us to test how a system will perform under different conditions and make adjustments to improve it without building the real thing first.

Models, mental models, and biases

Because most models (programs, diagrams, and spreadsheets) start as mental models rooted in our basic understanding of the world around us, they are vulnerable to all sorts of biases based on past experiences, culture, knowledge of history, professional training, etc.

Translating your mental models into mathematical ones allows you to become more objective and empirical about the logic and the input variables most affecting the desired outcome, thus mitigating the allure and traps posed by biases such as the illusion of control, the illusion of validity, illusory correlation, and so on.[10] If you look up decision-making biases on Wikipedia, there are 125 aggregated primarily in the areas of technology, business, and psychology. Since we don't want to boil the ocean, we have selected 25 of the most important biases that a decision superhero should be aware of, including examples and sources. Though we don't cover all of these biases in the book, we touch on most of them. *As you read the book, identify the bias we are tackling.*

TABLE 7: BIAS AWARENESS

Bias and Source	Description
Agent Detection: Heider, F. (1958). The Psychology of Interpersonal Relations.	The inclination to presume the purposeful intervention of a sentient or intelligent agent. For example, believing that deliberate actions rather than natural variability cause unusual weather patterns.

---

[10] (Makrdakis, Hogarth, & Gaba, 2009),

Bias and Source	Description
Ambiguity Effect: Ellsberg, D. (1961). "Risk, Ambiguity, and the Savage Axioms." Quarterly Journal of Economics.	The tendency to avoid options for which the probability of a favorable outcome is unknown. For example, preferring a familiar restaurant over a new one because you're unsure if you'll like the new place.
Anchoring or Focalism: Tversky, A., & Kahneman, D. (1974). "Judgment under Uncertainty: Heuristics and Biases." Science.	The tendency to rely too heavily, or "anchor," on one trait or piece of information when making decisions (usually the first piece of information). For example, a real estate agent shows you an overpriced house first to make other houses seem more reasonably priced.
Automation/Computer Bias: Parasuraman, R., & Manzey, D. H. (2010). "Complacency and Bias in Human Use of Automation: An Attentional Integration." Human Factors.	The tendency to depend excessively on automated systems, which can lead to erroneous automated information overriding correct decisions. For example, trusting GPS directions even when they lead you to an incorrect or dangerous route.
Availability Heuristic: Tversky, A., & Kahneman, D. (1973). "Availability: A Heuristic for Judging Frequency and Probability." Cognitive Psychology.	The tendency to overestimate the likelihood of events with greater "availability" in memory, influenced by recency or emotional impact. For example, overestimating the risk of shark attacks because of frequent media coverage.

Bias and Source	Description
Base Rate Fallacy or Base Rate Neglect: Bar-Hillel, M. (1980). "The Base-Rate Fallacy in Probability Judgments." Acta Psychologica.	The tendency to ignore general information and focus on specific case information, even when general information is more important. For example, ignoring statistical data about disease prevalence and focusing on a personal anecdote when evaluating health risks.
Belief Bias: Evans, J. St. B. T., & Curtis-Holmes, J. (2005). "Rapid Responding Increases Belief Bias: Evidence for the Dual-Process Theory of Reasoning." Thinking & Reasoning.	An effect where someone's evaluation of the logical strength of an argument is biased by the believability of the conclusion. For example, accepting a poorly constructed argument if it aligns with your existing beliefs.
Berkson's Paradox: Berkson, J. (1946). "Limitations of the Application of Fourfold Table Analysis to Hospital Data." Biometrics Bulletin.	The tendency to misinterpret statistical experiments involving conditional probabilities. For example, misunderstanding the relationship between two variables in a medical study due to selection bias.
Clustering Illusion: Gilovich, T., Vallone, R., & Tversky, A. (1985). "The Hot Hand in Basketball: On the Misperception of Random Sequences." Cognitive Psychology.	The tendency to overestimate the importance of small runs, streaks, or clusters in large samples of random data (seeing phantom patterns). For example, believing that a stock market pattern predicts future movements when it's actually random or believing a basketball player is "on fire" and will keep scoring because they've made several consecutive shots.

Bias and Source	Description
Confirmation Bias: Nickerson, R. S. (1998). "Confirmation Bias: A Ubiquitous Phenomenon in Many Guises." Review of General Psychology.	The tendency to search for, interpret, focus on, and remember information in a way that confirms one's preconceptions. For example, only reading articles that support your opinion on climate change while ignoring opposing viewpoints.
Dunning-Kruger Effect: Kruger, J., & Dunning, D. (1999). "Unskilled and Unaware of It: How Difficulties in Recognizing One's Own Incompetence Lead to Inflated Self-Assessments." Journal of Personality and Social Psychology.	The tendency for unskilled individuals to overestimate their own ability and for experts to underestimate their own ability. For example, a novice chess player overestimating their skills compared to more experienced players.
Framing Effect: Tversky, A., & Kahneman, D. (1981). "The Framing of Decisions and the Psychology of Choice." Science.	Drawing different conclusions from the same information depending on how that information is presented. For example, preferring a surgery with a "90% survival rate" over one with a "10% mortality rate," even though they are equivalent.
Gambler's Fallacy: Tversky, A., & Kahneman, D. (1971). "Belief in the Law of Small Numbers." Psychological Bulletin.	The tendency to think that future probabilities are altered by past events, when, in reality, they are unchanged. For example, believing that a coin toss is more likely to land on tails after several heads in a row.
Illusion of Control: Langer, E. J. (1975). "The Illusion of Control." Journal of Personality and Social Psychology.	The tendency to overestimate one's degree of influence over external events. For example, thinking you can control the outcome of a lottery by choosing "lucky" numbers.

Bias and Source	Description
Illusion of Validity: Tversky, A., & Kahneman, D. (1973). "Availability: A Heuristic for Judging Frequency and Probability." Cognitive Psychology.	Believing that one's judgments are accurate, especially when available information is consistent or inter-correlated. For example, trusting your stock market predictions because they align with past trends, despite the market's inherent unpredictability.
Illusory Correlation: Chapman, L. J., & Chapman, J. P. (1967). "Illusory Correlation as an Obstacle to the Use of Valid Psychodiagnostic Signs." Journal of Abnormal Psychology.	Inaccurately perceiving a relationship between two unrelated events. For example, believing that wearing a certain shirt causes your team to win games.
Information Bias: Baron, J., Beattie, J., & Hershey, J. C. (1988). "Heuristics and Biases in Diagnostic Reasoning: II. Congruence, Information, and Certainty." Organizational Behavior and Human Decision Processes.	The tendency to seek information even when it cannot affect action. For example, ordering additional medical tests despite knowing they won't change the treatment plan.
Neglect of Probability: Kahneman, D., & Tversky, A. (1979). "Prospect Theory: An Analysis of Decision under Risk." Econometrica.	The tendency to disregard probability when making a decision under uncertainty. For example, ignoring the low probability of a plane crash and deciding not to fly out of fear.
Normalcy Bias: Greene, R. R., & Greene, D. G. (2009). "Survival: What Firefighters Can Teach Us About Resilience." American Psychological Association.	The refusal to plan for, or react to, a disaster that has never happened before. For example, failing to prepare for a major earthquake because none have occurred in recent memory.

Bias and Source	Description
Overconfidence Effect: Fischhoff, B., Slovic, P., & Lichtenstein, S. (1977). "Knowing with Certainty: The Appropriateness of Extreme Confidence." Journal of Experimental Psychology: Human Perception and Performance.	Excessive confidence in one's answers to questions. For example, being sure that you aced a test, only to find out you did poorly.
Planning Fallacy: Buehler, R., Griffin, D., & Ross, M. (1994). "Exploring the "Planning Fallacy": Why People Underestimate Their Task Completion Times." Journal of Personality and Social Psychology.	The tendency to underestimate task-completion times. For example, underestimating how long it will take to write a report, leading to missed deadlines.
Survivorship Bias: Harris, S. (2013). The Survivorship Bias. Originally coined by Abraham Wald.	Concentrating on the people or things that "survived" a process and overlooking those that didn't. For example, focusing on successful entrepreneurs and ignoring the numerous failed startups when assessing business risks.
Zero-Risk Bias: Slovic, P. (1987). "Perception of Risk." Science.	Preference for reducing a small risk to zero over a greater reduction in a larger risk. For example, opting for a treatment that eliminates a minor health risk while ignoring a more significant, untreated condition.
Zero-Sum Bias: Meegan, D. V. (2010). "Zero-Sum Bias: Perceived Competition Despite Unlimited Resources." Frontiers in Psychology.	A bias where a situation is incorrectly perceived as a zero-sum game (one person's gain is another's loss). For example, believing that increased immigration must result in fewer jobs for native citizens, ignoring the potential for job creation.

To compensate for the biases that a decision superhero can and will encounter on a day-to-day basis, practitioners have successfully used different approaches to move from a mental model to a mathematical one. Let's check out the main approaches used to translate ideas and data into insights, and keep biases in check.

Different business modeling approaches

As you can imagine, different disciplines have different approaches and methods for rendering ideas into mathematics. Let's cover a few, including systems analysis, financial modeling, economic modeling, project scheduling, supply chain modeling, and industrial processes.

Systems analysis

At its highest level of abstraction, a **system** *is a collection of independent objects that are linked together and produce an identifiable outcome. Systems analysis is the process of studying, documenting, and optimizing a system toward some objective function.*

From a practical point of view, system analysis usually involves the indirect study of the target system through models of all or part of that system to identify patterns, relationships, and behaviors. The purpose of systems analysis may be to design a new system or improve an existing one based on some necessity. It may involve changing or redesigning the system itself.

Financial modeling

Financial modeling's objective is to give *a good enough view into the future that allows leadership to test operating scenarios and make strategic decisions.* Financial modeling aims to support decisions impacting a firm's sustainability, shorter-term profit, and shareholder expectations.

Financial modeling uses lots of standard decision-making tools and models (for example, M&A Scenarios, proforma financials, sales forecasts, discounted cash flows, total shareholder return, CAPEX project estimates, and so on). The primary analytical methods employed are (but are not limited to) time series forecasting, resource allocation and optimization, Monte Carlo simulation, and sensitivity analysis to account for factors that most impact the outputs of interest (ROI, Margins, Cashflow, NPV, and so on).

Economic modeling

Economic modeling looks at historical data on supply and demand and incorporates market data to predict the impact of social-economic trends and policy decisions. Economic models can consider entire markets to analyze supply/demand, forecast GDP and money theory, predict commodity prices, or anticipate the impact of policy changes. Similar to financial modeling, analytical methods employed are (but are not limited to) time series forecasting, resource allocation and optimization, Monte Carlo simulation, and sensitivity analysis to account for factors that most impact the outputs of interest.

In contrast to classic econometric models, some economic models are very specific in scope and can be summarized by algorithms such as the Black-Scholes equation.

Project risk analysis (schedules and cost estimates)

Project scheduling is a discipline where an analyst identifies tasks to accomplish, their dependencies, and their forecasted duration and organizes them logically in a structured project plan. Classical methods include the PERT (Program Evaluation and Review Technique) and CPM (critical path method), which were both developed in the late 1950s by the RAND Corporation and the US Navy to shorten the development life cycle of the Polaris missiles.

Most projects, especially the really big ones, use readily available tools such as MS Project or Primavera P6, which are designed to model their project plans and assess how much time and what kind of human capital resources are required for the project using a combination of PERT and CPM methods. Sadly, most of the analysis stops here.

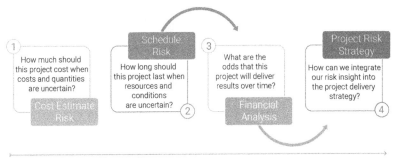

Risk Enabled Project Planning Process

Figure 33: Risk-enabled planning process.

For those managers and leaders who DO NOT want to be caught off-guard, an entire class of tools destined to simulate and assess the probabilities of success of either a cost estimate or work schedules has emerged. Using simulation, these tools stress test a project schedule and see which tasks contribute the most to making a project late and which resources are most likely to cost more. When working with project risk analysis tools, schedules and cost estimates are given in terms of ranges and probabilities.

Discrete event modeling (supply chain, manufacturing, and other physical systems)

Discrete Event Simulation (DES) is used to model complex systems as a process or sequence of well-defined events. *Each event occurs at a particular instant in time and marks a change of state in the system.* We can successfully model production systems, business processes, and traffic studies with DES. Just ask FORD!

Modeling requires the right attitude

The relationship between an analyst's humility and accuracy is proportional.

Insight is better than specific answers

The famous statistician George Box once said that *"all models are wrong, some are useful."* What did George mean by this? Essentially, he is saying that the world is too complicated to put into a simplified mathematical form, but doing so still gives insight you did not have before building a model.

What is the difference between insight and information? Insight is understanding, and information is a set of data points that need to be interpreted for insight.

Former general and United States president Eisenhower agreed with George (though we don't know if they ever spoke) and famously said, *"Planning is everything and plans are nothing."* By doing the planning, simulations, training drills, and thought experiments, you are better prepared in your mind for any eventuality—a point equally made almost 400 years earlier by Japan's fiercest samurai, Miyamoto Musashi, in the "Book of the Five Rings."

*To this effect, because there are so many unknowns and uncertain quantities in any model or plan, any actual plan has little or no chance of being implemented as is. **Therefore, developing insight on responding to different eventualities as they arise becomes the most crucial skill.***

Don't get caught up in the weeds of your own mind

We sometimes invest so much time and effort into the minutia of a model that we lose perspective that it can be wrong (hence why getting a second pair of eyes to do a sanity check is critical). With enough successes under your belt, you might start being recognized as a Rockstar who can do no wrong with math. If you start thinking that you are beyond reproach or amazing at everything you do, then at this point, you are in the very dangerous territory of denial (and not The Nile), justification, and tunnel vision.

> *"Don't believe your own hype."*–Origin unknown

Like the fox, the solution is to stay grounded and open to new information and ideas. Being challenged is not always easy, but if your thinking and source data can back it up, it's a worthy exercise in defending ideas. On the other hand, **if you realize that what you are defending is wrong, then just admit it and move on.** *No one will remember that you changed your mind in light of new information, but they will remember the one who argued a false point to death to save face.*

The preceding is the most direct advice I can give to understand the necessity of intellectual honesty, propensity for clarity and precision, and healthy skepticism in modeling. Remember them well because a great analyst is humble [about results] and strives to maintain objectivity on the data and the methods.

Modeling is an application of the scientific method

Eventually, by adding simulation to a model, you can also test your mathematical hypothesis (what you think the answer might look like) using brute-force numerical methods as well as see whether the starting assumptions were valid or require further refinement and investigation. Good modeling is, in fact, a simplified application of the scientific method—a process of testing ideas (hypotheses) by adding new information and proposing alternate hypotheses when old ones [ideas] prove wrong. By cycling through the process multiple times, the modeler can derive valuable knowledge on the dynamics of a system. Actually, this process can be more valuable than the model's results because a deeper understanding of the dynamics can often translate into better decision-making under uncertainty.

The scientific method is nothing new and some trace it back all the way to the 17th century. The scientific method is based on methodical skepticism and tries to account for cognitive assumptions and biases when seeking an answer to a question. In the classic Business Research Methods (Cooper & Schindler, 2003), they propose six general characteristics of the scientific method as applied to management questions:

- The direct observation of phenomena (measurements and instrumentation to collect data)
- Clearly defined variables, methods, and procedures
- Empirically testable hypotheses
- The ability to rule out trivial hypotheses
- Statistical rather than linguistic justification of conclusions
- A self-correcting process

By going through the preceding list, we recognize that these steps are almost implicit in any business analysis. It is worth mentioning that iteration and self-correction (a systems approach) equally support our statement that modeling is an iterative process.

What kind of model are you trying to build?

Various model types/dimensions exist. We shall cover several important classifications for decision and risk analysis disciplines. The following figure is a series of considerations the modeler needs to make to establish the design characteristics of their analytics project.

Of course, it would make no sense to go any further if we were dealing with a qualitative analysis here.

Figure 34: Questions to answer when scoping a model.

Given that we are working on quantitative models, we shall assume that is the answer to the first question so we can continue exploring the characteristics of quantitative models.

Quantitative versus qualitative models

As the names imply, quantitative analysis looks at quantifiable values (hard numbers), while qualitative models seek to take abstract concepts such as experiential data and translate them into numbers we can monitor or manage (e.g., customer satisfaction index). The primary tools for qualitative data are surveys, interviews, and testimonies, while historical data is the basis for most quantitative analysis.

Descriptive models

Descriptive models are the basis for all further modeling. Descriptive models are simply a description in mathematical form, usually in deterministic terms. *They describe what is.* A good example is an accounting report such as an income statement or perhaps a sales report that gives you the state at a specific point in time with clear and understood calculations. In the case of an income statement, you can test assumptions by playing with the input variables.

Another, geekier example is correlation. It's a descriptive measure that seeks to explain the relationship between two variables, for example, marketing expenses versus sales performance.

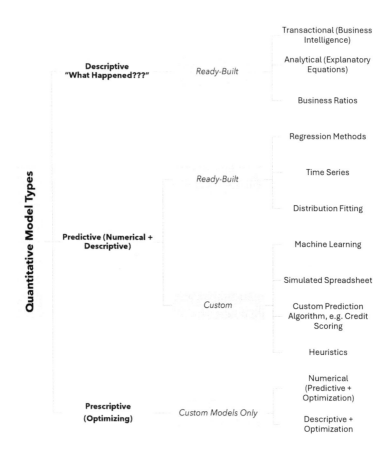

Figure 35: Modeling techniques by method.

Predictive models

Predictive models come in different forms. We shall cover the two most popular ones, which are machine learning and simulation. Machine learning prediction models look at historical data and produce logic statements known as business rules. Essentially, a rule is a set of conditions that, if met, will produce a predictable outcome. If a model structure is unknown, machine learning can also give clues to the modeler of unobserved underlying relationships. For example, when analyzing credit risk, you may consider that age, annual income, and marital status are good loan collection predictors. Certain combinations of these predictors will dictate the loan has a good chance of being fully collected, while others would predict default.

Another approach to predictive modeling is to build a descriptive model first using expert input and then replacing key inputs with probability distributions—more specifically, the model first and the data after. Through this approach, you can incorporate uncertainty in the inputs and assess their impacts on the outcomes, also providing insight into key influencers on the target outcome.

Prescriptive models

Prescriptive models (also known as optimizing models) also come in several flavors, but as in predictive modeling, prescriptive models start with a descriptive model. What is central to optimization models are inputs known as decision variables. Decision variables are things in a model under the modeler's control. For example, how many people should staff the sales desk on a Wednesday afternoon when business is historically slower versus a Monday morning when business is brisk? Or how many widgets and confabulators should we buy based on a contractual service level?

The two main ways of optimizing are *analytical* and *numerical*. Analytical optimization reduces your model to an algebraic equation and solves for a specific outcome. While analytical optimization is cool and fast, some problems do not lend themselves to those methods. For these problems, we have a brute-force numerical optimization technique known as stochastic optimization or Monte Carlo optimization. This approach takes a predictive model (descriptive model + simulation) and incorporates decision variables into it so you can optimize against an uncertain outcome (see the following figure).

Though some models are inherently probabilistic, this is not always the case or even necessary, given the problem to solve. If we are doing descriptive work and looking at historical data, there is no probability in history—there is only certainty. On the other hand, if we are forecasting or trying to assign a probability to an outcome, then probabilistic modeling is needed.

Deterministic versus probabilistic

A deterministic model is one where you can calculate the output precisely given a specific set of inputs. For example, if Revenue is $3M and Expenses are $2M, then per our model [Profit = Revenue-Expenses] our Profit IS $1M.

Probabilistic models often leverage the same logic as their deterministic counterparts but allow you to use random variables (probability distributions) as inputs through the model and view the result as a probability distribution where you can assess the likelihoods of various scenarios.

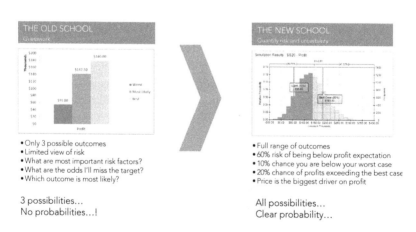

Figure 37: Classic scenario analysis versus simulation using the same inputs.

Analytical versus numerical

This is perhaps an old-school distinction. Analytical models are based on algebraic modeling using a symbolic form. That means you distill the situation into equations and solve for x. Period. Numerical models also require an equation, but they can be far simpler. You push thousands of numbers or scenarios in on one side and record the output on the other using the brute force of modern computers. Then, by looking at the aggregate outputs, you can see which variables correlate and get the "area under the curve" at specific confidence intervals. This can be done using analytical methods, but you will probably end up with some crazy integral calculations.

A worthy piece of advice is never to reinvent the wheel or automate the phonograph. If tools, functions, or models already exist for a particular problem, use them. This is equally true when considering a make-or-buy decision for all analytics endeavors.

Ready-built versus custom (functions versus programs)

A ready-built model is one where the algorithm is generally understood the same way by all who use it, such as sums, regression, probability distributions, NPV (Net Present Value of future cashflows), or business ratios. You need to plug in your numbers and know what to expect on the other end. Most functions in Microsoft Excel and those available in most Matlab, Python, or Julia packages can be defined as ready-built. The intent of standardizing code into ready-built functions is that it considerably streamlines development and avoids the need to include endless lines in each project just for standard functionality.

Custom models are unique to your project. *They usually take the form of custom libraries, algorithms, User Defined Functions, entire scripts, or programs.* These models are built with a specific application in mind and are *usually constructed with many functions.* The most basic example is the common spreadsheet. Your spreadsheet contains a model, a series of inputs, and a bunch of functions (e.g., `SUM()`, `AVERAGE()`, and `COUNT()`) that, strung together, will provide you with the intended answer (output).

Key takeaways

- Decision Superheroes trade in models, and successful models result in good decisions.

- Every model must answer a question and have the three universal components: Inputs, Outputs, and Model.

- Different model types and methods allow a decision superhero to frame and synthesize different mental models and philosophical approaches, such as financial modeling, systems modeling, economic modeling, DES, and project risk analysis.

- Most modeling approaches share some or all of the same mathematical techniques, and having a clear vision of which tools are best for solving the problem makes a big difference to the quality of the final outcome.

- Models can be descriptive, predictive, or prescriptive. They can be probabilistic or deterministic, ready-built or custom.

- Models have common types but are also unique to their circumstances. Put simply, models are like stories—they have universal structures and storylines, but the characters and the plot specifics are what make two similar plot lines different.

In the next chapter, we will explore the economics of modeling and when it makes sense to automate or take the time to develop a bespoke solution.

Six

Selecting the Right Modeling Process

Every decision superhero knows that not all problems and decisions need the same level of analysis. Knowing when to code a model and when to use a spreadsheet is often a question of habit for many, and it might be worth rethinking that choice. In this chapter, we seek to give an economic framework for deciding when to code a solution rather than develop a spreadsheet. Because selecting the right approach directly impacts the cost and speed of getting an answer, it becomes important to have a framework for evaluating which strategy is best.

This chapter proposes four development strategies based on the frequency and utility of models. Selecting the right strategy depends on setting and measuring objectives while keeping in mind a model's performance and end goal. We will explore how the ideas of decision value/utility and frequency directly impact the appropriate time invested in developing a solution and updating it with new values.

In addition to the conceptual framework, we have a practical example to compare the benefits of developing a spreadsheet and a coded model. We will see how to calculate break-even and solution costs over time by applying a simple rational comparison approach.

Knowing when it is *appropriate to code and when it is better to use other tools/environments for rapid and one-off decisions* is another important superpower—so let's dig in.

Why selecting the right approach matters

Thomas Edison said it best, "*Vision without execution is hallucination*," because the link between knowing what's best to do and its action is clear. Edison was ahead of his time because business execution is often ranked as the number 1 critical success factor by several modern authoritative works on the subject, such as Nohria's seminal 2003 Harvard Business Review article and the management classic *The Boundaryless Organization* (Ashkenas, Ulrich, & Kerr, 1995). The other critical success factor for modern business is velocity/speed. Organizations that deliver faster have the shortest cycle time for analysis, answer concerns faster, understand market patterns quicker, capitalize on opportunities more rapidly, and generally outperform their peers. By focusing on these two critical success factors, smaller firms can derive a competitive advantage over larger firms that enjoy more resources and scale (McNeilly, 2012).

Setting a goal for your analytics model

An analytics model can only deliver faster decisions or better ones,[11] including a combination thereof. Knowing which of these ideals is your primary goal allows you to determine a development strategy, what kind of tools/platforms to use, and how many resources to allocate. From a practical point of view, analytics provides a pathway to answering questions of all kinds, for example, *How much should I*

---

[11] This applies also to the organizational process more broadly, see chapter 7).

make of A versus B?, *What day is best to advertise on the web?*, *Where should I place my next store?*, *Should I build a new plant?*, and so on. Each of these questions has a different value and utility and requires answers—usually fast.

Getting the problem-solving approach and development strategy right will translate to faster cycle times and more opportunities for further insight. *Most business questions involve trade-offs or opportunity costs, so it is important to remember that a dollar spent solving one problem will not be spent on another.* Ultimately, success lies in matching the most efficient and effective strategy with action to achieve results better than the other guy/gal.

Considering what is good enough is often based on what it will cost and how long it will take to conduct an analysis. Because we do not live in an unconstrained environment, we must consider financial and scheduling constraints when developing an Excel analysis or coding a model.

What are the properties of the decisions we make?

Some decisions are easy while others take more work, and some decisions happen rarely while others every day. The combination of frequency and utility will dictate which approaches to use and how much effort to invest to get the right answer.

Here are some basic questions to ponder while scoping both the question and the modeling approach:

- *Define the value/benefit of the decision*: When we talk about decision utility, we mean that decisions and priorities are made based on the values and economics of your business. For example, payback, various ROI metrics (return on investment), and process capability improvement are operational and financial metrics to compare a decision's relative value. Selecting which metric(s) is the most appropriate measure of success should be considered from the onset.

- *Define the cost of the decision*: This is necessary for applying utility theory and basically making trade-offs based on bang for the buck.

- *Time horizon for taking action*: You have lots of time or little time to decide. This will dictate how much effort is required and how granular your analysis needs to be.

- *Frequency of decisions*: If a decision happens regularly, how often do you have to make it? The more frequently you make a decision, the better the ROI can be on your analytics project.

The following figure shows simple examples of each decision type classified using the preceding questions:

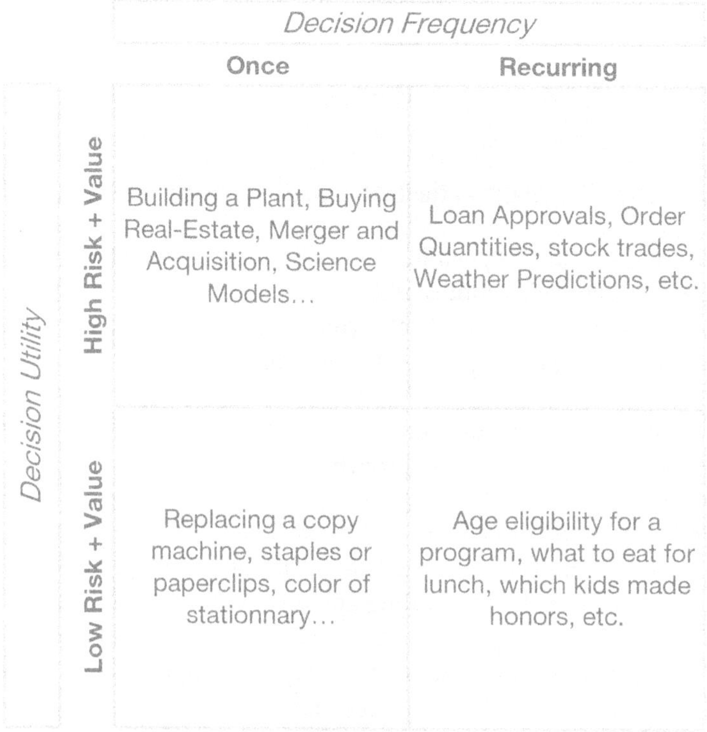

		Decision Frequency	
		Once	**Recurring**
Decision Utility	High Risk + Value	Building a Plant, Buying Real-Estate, Merger and Acquisition, Science Models…	Loan Approvals, Order Quantities, stock trades, Weather Predictions, etc.
	Low Risk + Value	Replacing a copy machine, staples or paperclips, color of stationnary…	Age eligibility for a program, what to eat for lunch, which kids made honors, etc.

Figure 38: Examples of decisions by utility and frequency.

What kind of decisions are you planning to make?

A very important consideration is which strategy we want to employ to make decisions. Decision frequency and utility (value) are the two major dimensions to consider. In the following figure, using these two major dimensions derives four major strategies that will entail very different approaches to solving the problem:

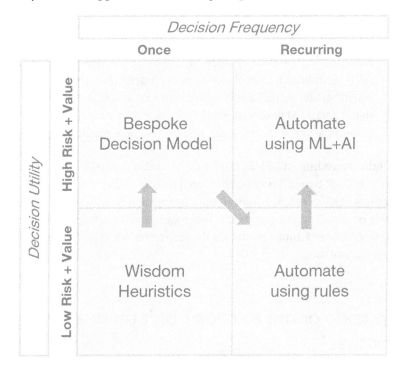

Figure 39: What kind of model should I develop?

Here is a description of the major strategies. For practical reasons, we have put the modeling strategies in order of effort/cost/benefit:

1. *Wisdom heuristics* (your brain) are basically using experience to make decisions.

2. *Bespoke decision models* are built when a decision is rare or unique but greatly impacts the firm. For example, mergers and

acquisitions, opening up a new national office, or large capital expenditures such as building a new plant Capital expenditure (CAPEX). Once the decision is made, the model is filed away as support material.

3. *Automating using rules* is about taking formal classifications, such as age or number of units bought, and creating a business rule to be applied automatically to the data during processing. This is usually a coded solution.

4. *Advanced automation of rules and models* can make use of data APIs, simulation, optimization, pattern detection, machine learning, and artificial intelligence on vast amounts of data to make decisions and recommendations automatically. These systems, when well designed, will evolve their predictive ability.

All the preceding strategies can be developed as coded solutions, but only the first two can also be envisaged as spreadsheets. Once a strategy has been selected, it may already be implicit whether you have to code or not. If starting appears challenging from a strategic point of view, consider a more operational perspective focused on the most appropriate tool.

To code or not to code? Isn't there a tool for that?

When working with a one-off decision, consider just doing a quick Microsoft Excel model because there is no need to run the model in the future and invest all that time coding it. On the other hand, as your understanding of the problem's frequency and utility evolves, it may become apparent that this task needs to be automated and potentially evolve to use more sophisticated methods and tools.

Figure 40 shows, with arrows, that there is a logical, evolutionary life cycle to consider when evaluating which strategy is best.

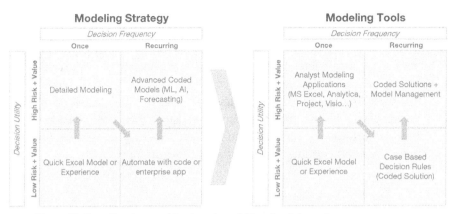

Figure 40: From Excel to machine learning-picking the right tools.

As a model demonstrates value in the early phases of development, you can decide to build out more strategic aspects to improve accuracy. The other consideration when developing a model in greater detail is whether we need to run this model regularly. Sometimes, a one-off model can identify a hidden need and become a model prototype we can automate. For example, we can build a time series forecast in Excel and manually load into the ERP system, or run as a script that pushes the results directly. How many important and strategic things are done in loosely strung-together spreadsheets is amazing. **Any spreadsheet that gets updated with some frequency is an opportunity to automate.**

If the decision-making model is more sophisticated, such as credit scoring or identifying potential prospects for online marketing, we deal with highly repetitive decisions with unique characteristics. Given that we need to process unique and frequent decisions, we may elect to implement more sophisticated machine learning-based models to account for frequency and personalization. The key takeaway from the preceding figure is that there is an evolutionary life cycle to respect when selecting a toolset. Once we hit the sweet spot, we stop evolving the strategic aspects of the model and start focusing on its specifics. Whenever possible, go for the economy of scope. An *economy of scope* in analytic terms means that developing one model or model component reduces costs and errors when developing other interrelated or dependent models.

One-off/low-risk/low-value decisions ($)

If you are only making a decision once and getting it wrong will not have a major impact, go for what is fast and generally reliable, such as an Excel model, a pen and paper, or gut instinct. When it comes to simple one-offs, though you may make less optimal decisions on the spot, you will win out over time because of speed or, better yet, *time-to-decision*. For instance, you can save 48% on your new equipment if you buy it today instead of a month from now.

One-off/high-risk/high-value decisions ($$$)

Big decisions usually carry big price tags and lots of uncertainty about profitability, cost, or delivery dates. When planning to spend a billion dollars on improving a highway or a paltry 100 million on building a processing plant, you tend to want the best forecast on what it will cost and whether the investment makes sense. It is typical for organizations to take a year and a team of analysts to get it right (and for good reason). Technically, the more information and analysis you conduct, the less wrong you will be. For this, the Monte Carlo simulation has proven to be an extremely useful tool to be less wrong. Remember that the value of the models used to make this decision goes to zero as soon as the project is delivered, so it rarely makes sense to code it unless your domain of work requires it.

High-frequency/low-risk/low-value decisions ($$)

On the other hand, when a decision gets made over and over again, well-tested software vendors or coded solutions are the way to go for repeatability and speed of execution (time to decision). We can configure many of these decision models as rule sets or simple algorithms in existing business systems. Economies of scale are what we seek when automating simple decisions.

High-frequency/high-risk/high-value decisions ($$$)

High-frequency trading, airplane scheduling, supply chain, portfolios, credit monitoring, and operational risk are all areas where making optimal decisions at the right time can make the difference between an organization's life and death. At the very least, the difference between making money and not making money.

There are entire sectors where having a competitive advantage in both speed and quality of methods is the driving force behind their success. Banks, retailers, institutional investors, pension funds, governments, and any major organization impacting our day-to-day lives have implemented advanced models blending forecasting, machine learning, artificial intelligence, and the simulation of outcomes. We build these models as bespoke coded solutions (often quite proprietary) that run at blinding speeds and produce the deep insight/analytics required to manage their risk at a reasonable and/or regulated level.

Can a model evolve from one type to the next?

Given that coded solutions take more time and consequently cost more money to develop than bespoke models, the ROI on the time saved accrues every time coded models are run. For this reason, we may find that a model that started as a one-off is, in fact, a model we should code and run regularly. All models start as a simple model and then a decision is made about whether to conduct further analysis. This analysis usually serves as a proof of concept or as a systems analysis destined for either developing code or configuring an existing business system.

Developing models is an iterative process

Because modeling is an iterative process, Agile methods lend themselves well to researching and developing analytics solutions because they focus on delivering usable models in short development cycles (Rogers & Thompson, 2017). Getting funding or backing for a project will often require early results and validation from business stakeholders.

Another key part of the iterative process is to keep the subject matter experts and the client in the loop throughout the process to validate both the methodology and the soundness of the results. In the absence of a subject matter expert, you should model several other approaches to see whether you get within a certain ballpark of answers.

Triangulating answers and testing ideas before doing the real work

Different people have different go-to methodologies or tools to get answers. If you are an engineer, you probably have a certain set of formulas and templates from which to work. If you are an accountant, spreadsheets and accounting software are what you reach for to get the job done.

For example, if you are working in the area of financial modeling, you may calculate the value of an option from time to time. Simulation offers many advantages over conventional methods, but you may need to ensure your model produces sound results. Many years ago, one of the best pieces of advice I got regarding this problem was to triangulate an answer. Usually, an analytic or closed-form solution, such as **Black-Scholes**,[12] can be used to see whether your simulated option pricing model results are in the right order of magnitude and the correct ballpark of results. If you can access other methodologies, such as binomial lattice models, you could also de-

---

[12] The Black-Sholes formula is used to analytically evaluate the price of an option. The authors won a Nobel prize for the effort.

rive an answer that way and see whether your simulation results are also concordant.

People like to test solution ideas in different ways based on what they know, which can give them a good enough answer to go or not. If you intend to write a valid and satisfying model, you need to establish a starting point. Before committing to a development strategy or a particular solution, we start with writing a toy model in Excel or a few lines of code in Julia to see whether the idea makes sense or is feasible. Mostly, I (Eric) recommend starting with a proof-of-concept model in Excel because it is easy to share results and validate an algorithm with a wider group of people.

Usually, these proof-of-concept models take the form of Primitive Programs or, as Sam Savage calls them, *Paper Airplanes*. They are simple programs or spreadsheets designed to solve one problem and then be thrown away. The model developer may develop a primitive program (paper airplane) to demonstrate a particular algorithm or solution method (McNitt, 1983). As Sam puts it, Excel (plus a simulation) is an ideal environment for developing, testing, and collaborating on ideas, or as he puts it—*making throw-away paper airplanes.*

Paper airplanes are valuable for the following reasons:

- They allow you to figure out what you can and can't do

- They give boundaries to your problem's potential solution

- You can fail fast and move on to the next idea more quickly

- They allow you to explain a problem simply because they are small

- They can start as a diagram or a spreadsheet and are easily translated into code

Once the basic model is ready to go and we assess both the frequency and utility, we can estimate the total cost of ownership of both a spreadsheet solution and the coded solution over time.

The economics of code versus spreadsheets[13]

In 2019, we published a benchmarking study on Julia and used Markowitz's *Mean Variance Portfolio Optimization* model to compare solution performance and accuracy. In this section, we have a break-even analysis of when you should consider coding a solution versus developing it as a spreadsheet. The complete paper and results are available on GitHub at https://github.com/etorkia/NFS2019.

Both the Julia and Excel versions of our model start at the point where the data was downloaded and the log returns pre-calculated. Though a subsequent version of our model could download and process the financial stock ticker data, we elected not to use it since the Excel model did not download the stock data automatically—we wanted to keep the playing field level.

To build a simulated Markowitz Model, you need to perform the following steps:

1. Download the data for all the tickers using CSV (2min/ticker).

2. Consolidate asset/ticker data into one table for processing.

3. Calculate the log returns for each asset.

4. Calculate the correlation and covariance matrices.

5. Update tables and matrices to reflect the new number of assets.

6. Run the model.

Our timing test focused on adding one asset class or ticker to the model and re-running (the preceding steps 4-6). In the case of a port-

---

[13] Example taken from Need for Speed 2019 study I did in 2019 comparing Excel Monte-Carlo add-ins to Julia and R.

folio model we need to run many times, the answer is a no-brainer. But in reality, it's more subtle.

Under these circumstances, the economics and ROI are clear with a simple break-even calculation stating that running this model more than once a week would justify the investment in a coded solution that is less error-prone upon updates, as suggested by the benchmark data in Table 8.

TABLE 8: RUN AND DEVELOPMENT BENCHMARKS OF CODING A MODEL VERSUS DEVELOPING A SPREADSHEET.[14]

Metric	Julia	Excel + @RISK
Development Time	15 hrs	1.5 hrs
Update Time	1 min	15 min
Run Time	151 ms (100k)	58,900 ms (100k)
Updates/Hr	60x	4x
Max Sim/Hr*	24,000x	60x

Using the numbers provided in Table 1, we see in r that a coded solution with 15 hours of development time breaks even with Excel at 60 runs. However, to update the Excel model, you invariably touch the model's logic and structure, which can lead to correcting old errors while introducing new ones *because studies have suggested that the initial probability of a significant error in your Excel model is about 90%* (Harris, 2017). *On that basis, the update time is not being factored in*, which could significantly reduce the break-even. You also need to perform additional testing every time you update your Excel model and its data. In contrast, a coded and well-tested solution should not require more than a data update. This also mitigates the

---

[14] Tests run on a DELL Precision workstation T7600 using dual XEON E5-2620 2.0Ghz with 6 cores and 64gb RAM

risk of introducing new errors into the code, which avoids losing time diagnosing and testing results.

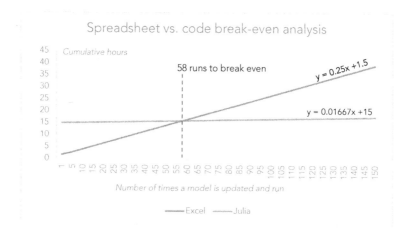

Figure 41: The economics of code versus spreadsheets.

Calculated using a simple linear model, Figure 41 presents the cumulative time for developing and running a spreadsheet model (gray) and a coded model (blue) using the values presented in Table 8. To set up the lines, we use the mx+b formula to calculate cumulative time in hours:

$$Cum.\,Time = (Updates \times Update\ Time) + Development\ Time$$

When comparing the cumulative runtimes (and commensurate billings), upon visual inspection, spreadsheets seem like a more cost-effective approach for anything requiring less than 60 runs within a year, all the while assuming minimal debugging and auditing.

On the other hand, if this model runs daily and for multiple clients or products, the coded solution allows it to run often with far less risk of errors and bugs due to copying or simple human distraction. Under these conditions, the coded solution would take 2.85 working months to pay for itself, notwithstanding the potential revenue gains from the massive reduction in cycle time and errors.

Granted, we can do a lot to automate Excel, but we are strictly comparing our models in their current state. To make the functionality of Excel as automated as our Julia version, the *Dev time* would be far closer to 15 hours than 1.5 hours due to the need to incorporate some VBA (Visual Basic for Applications, Microsoft's macro language).

If you develop an equivalent Excel VBA solution, you will never have the same raw power, speed, reliability, or scalability and for this reason **Julia (or Python) is the better option from day one for this type of model.**

In terms of accuracy, studies have proven that computers will outperform humans on repetitive decision tasks such as loan approvals and school admissions. The main reason for this is that coded computer algorithms don't have the inherent biases of humans and will apply rules and approvals consistently (Microsoft Corporation, 2018). If biases are built into your models by accident, usually a risk in machine-learned models, there will be biases in your decisions. The main point to remember is that a machine will consistently apply your decision criteria, well thought out or not.

Lastly, though you can segment your decision model development projects in many ways, remember that the utility and frequency of decisions are essential and should be at the root of any analytics budget/allocation strategy.

Key takeaways

- After defining the scope of the model, determine the best technology, methodology, and tools.

- Classifying decisions by relative value and frequency is usually a robust framework for selecting the modeling process and tools.

- Apply the simple framework for comparing the value of coding a solution or developing a spreadsheet model.

- Keeping in mind that models are an iterative process makes sure that important things are constantly being re-validated and provides the reassurance that if there are errors, we can find them quickly.

- Always try multiple recognized methods or find appropriate benchmarks that, with relatively little effort, confirm whether the results produced in the model make sense.

- Certain problem types can very efficiently be solved using analytic models.

- With an economic framework for selecting the appropriate development strategy, a firm can start developing a roadmap for making decision science a core competency.

The Decision Science Roadmap

Building an effective decision science solution requires domain experience and correctly targeting a question and translating it into a mathematical model that leads to actionable answers. Naturally, if you answered the right question and did the right analysis, effectively implementing these models in the form of code and business process changes will eventually lead to the results you are looking for.

Generally, having a framework or a vision for both a decision science project and workflow function can greatly accelerate adoption. In the following pages, the goal is to introduce you to some key definitions, concepts, and workflows that will allow you to both scope and plan your next initiative clearly and concisely.

Success is getting the results you want. In the following sections, we will cover the high-level decision science process and the key activities you need to include in business processes and your project planning. The main difference between a process and a project is that you can run a process many times, but a project only runs once. So, you can use the following to either design a business process or elaborate on your next decision science project successfully.

Do you want faster or better decisions?

 There are only three ways to get better outcomes in de-cision science: FASTER, BETTER, or a COMBINATION OF BOTH.

> *Decision-making is basically a problem where you need to select the best option among a predefined set of alternatives.*

Faster decisions

The speed and velocity of an organization's decision-making lifecycle are often cited as a critical capability and attributed to most of the world's most successful companies. *In most cases, faster means au-tomation.* From a decision science point of view, some decisions get made once or very rarely, while others are made repeatedly (daily, weekly, monthly, or quarterly). In some cases, some analysts (smart human automatons) run a few specific analyses in Excel because they are so time-consuming and onerous to manage. This is where an organization may seek to do some white-collar automation with the benefit of massive speed improvements, error reductions, and human capital cost savings. We propose a framework to analyze the value of automation in Chapter 10.

As a gentle piece of advice, it is always better to be the *"automator"* than the *"automatee."* So, if you ever recognize that you are perform-ing a repetitive analysis, be the first to propose a solution to auto-mate it. You will then become an *automator* and can automate where the needs arise, leaving you with many potential projects that will keep you relevant.

When a model goes from 8 hrs to 2 min to run

We transformed a spreadsheet-based simulation model into Julia code in a project with a major insurance company's consulting division. Originally, the spreadsheet model took four to eight hours to update and an additional 20 minutes to run 10,000 Monte Carlo Simulation trials. After migrating to Julia, we built the model in under 1.5 minutes and executed it in less than five seconds. This improvement significantly enhanced productivity, allowing analysts to perform multiple analyses daily, compared to the full day previously needed for a single analysis. From a profitability standpoint, the process that previously took eight hours now takes only two minutes, resulting in a substantial return on investment. This efficiency boosted business flexibility and allowed the firm to offer clients more frequent and cost-effective services, thereby increasing client satisfaction and decision-making agility.

Better decisions

Better decisions mean that your predictions lead to better decision outcomes more often. This is difficult to measure because you can make a good decision and have it go the wrong way or make a poor decision and have it go the right way. Nevertheless, it is worth investing time and effort to see if some techniques or methodologies deliver better results, and generally, this kind of R&D pays off.

The key challenge when designing for better decisions is not to get into a cycle of perfection-seeking, colloquially referred to as *paralysis by analysis*. One of the better strategies is to look at your design time in terms of the *Pareto power law*, which is the 80/20 rule. Through this lens, your goal is to focus on the 20% effort that will give you the 80% benefit upfront and then analyze how much of the remaining time you have to invest versus the performance gain you want. Tools and methodologies are available to assist in rightsizing further investments in your model, most notably the value of control and information. An interesting approach to the value of control is presented in the book *How to Measure Anything*, by Doug Hubbard.

Setting the analytical stage

The challenge with any generally accepted concept is that sometimes, we struggle to come up with a common definition because it seems so obvious. For both decision science and data science practitioners, it is important not to gloss over this very central distinction between exploratory analysis and prediction because it can lead to a lot of self-inflicted tail-chasing.

Exploratory analysis is to understand

As the name implies, exploratory analysis explores historical relationships and behaviors in the data to gain an insight and decision-making edge. The most important dimension of exploratory analysis is looking at historical data and relationships. That is why more advanced exploratory analysis can help you identify trends and changes over time that can lead you to better predictions. Furthermore, if the past is a good predictor of the future, then sophisticated fitting techniques (curves, probability distributions, time series, and so on) can be used to take historical data and trends and project them into the future when you get to the prediction phase of your process:

- A good exploratory analysis process will lead to three basic deliverables:

- An understanding of which variables are controllable and uncontrollable

- The behaviors of uncontrollable variables (modeled as probability distributions)

- The logic and strength of relationships between variables (usually modeled as functions and correlation)

Exploratory analysis can be conducted using many tools and can take many forms. Most companies analyze their data using Microsoft Excel to pull basic KPIs and chart the data. Some organizations opt for more centralized approaches, including tools such as Tableau or Microsoft Power BI, to present the historical data and KPIs to a broader audience. From an analytics maturity perspective, this is the first rung on the ladder. Conversely, many organizations elect to consider business intelligence, an automated exploratory analysis, as predictive analytics and consider the innovation job done. This is far from the truth because it is the business equivalent of driving a car on a highway by staring at the rear-view mirror.

Prediction for decisions

The goal of building predictive models is to make decisions. Period. We use predictive models because they are easier and cheaper to manipulate than their real-world counterparts or because it is impossible to manipulate or experiment with the real-world system. For example, models exist that predict what a man would weigh on Jupiter even though we really don't know. In fact, when we finally make it to Jupiter, we may realize that our estimate was totally wrong. Predictive analysis aims to have an anchor point against which we make decisions and eventually validate whether we are wrong or not in our assumptions. Sometimes, that validation can take a long time, frustrating many analysts who may skip or minimize validation efforts. This is not the best course of action.

Ultimately, the proof of a successful prediction is to see it happen in real life.

The decision science roadmap

In Chapter 1, we saw that the goal of the work is to support decision-making. Figure 42 shows the ideal flow, but there are iterative loops at every step, as required. For example, when you gather data, you may make observations that cause you to rethink your hypothesis or perhaps select another problem-solving approach. Things are never as clean in the real world as we would like. Thus, it's super important to be flexible by thoroughly understanding the process and where you are in the process.

The decision science process is broken into five major steps for practical reasons, such as fitting into other parts of the organization. Each step in the process has its own key deliverables associated with it that are necessary to move on to the next step in the process.

Each step has integration points with broader processes and systems such as IT, Strategy Planning, Finance, Operations, Project Management, CAPEX decisions, and so forth. For these reasons, the master process is simple and recognizable to all levels and groups of the organization.

The following is a high-level presentation of each step in the decision analysis roadmap.

1. **Defining a research question** while capturing all the nuances that influence that definition as well as set the criteria for selecting the best option (including the status quo).

2. **Collecting data, understanding the logic of the system**[15] under study, and developing assumptions/inputs to use in the model.

3. **Condense all the data, logic, and subject matter expertise**

---

[15] System is taken to mean a model consisting of mathematical relationships. An income statement (e.g. Revenue-Expenses = Profit) is both a model and a system.

into a mathematical form that allows for the testing of scenarios and inputs.

4. **Generate viable alternatives/options** (courses of actions) for the decision-maker to consider.

5. **Automate key models** so the decision-making insight is available when needed. It saves money too!

To give a sense of all the steps and flow for the roadmap, we have presented a summary of all five major steps. However, in subsequent chapters, we will delve into each major theme to highlight stories, lessons learned, best practices, and major considerations.

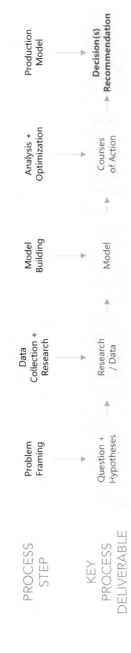

Figure 42: The Decision Science Process.

Problem framing process

Problem framing is the process of scoping and defining a decision. The process starts when we need to make, or perhaps a question is asked of you, or your boss says there is a problem with some aspect or dimension of the business. The next thing to consider is whether this "*question*" is amenable to finding a solution through an economically viable analysis. We often overlook it, but a critical aspect of the process is explicitly defining our hypotheses. That is, what we think the outcome will be. Doing so gives us a base case and an anchor point for further analysis into whether we should proceed. Do methodologies already exist that are accepted for the problem at hand? At this point, it is important to estimate whether the value of doing something exceeds the cost of the analysis. Otherwise, you may as well quit while you're ahead and find another, more valuable problem to solve for the organization.

The next step in the process is to frame the analysis in terms of what you are trying to find out and what you will do mathematically to get an answer. What do you think is going on? What are some of the elemental relationships?

Sometimes, a good starting point is to go to a whiteboard with a domain expert and try to identify the main elements of the model and the impacts when you change the system. What are the drivers you think will move the needle on results? From each of the important objectives, goals, or values you put forward in the decision model, you need metrics to assess the degree of achievement for each. When documenting your project, consider the variables you think need further analysis and data collection because this directly impacts cost, time to answer, and the accuracy of your solution.

Resource availability should not be shrugged off either when planning a decision science project. Often, company and subject matter experts are very solicited, and it can take some time before you sit down and interview them. This has a direct impact on your delivery date. If the project is time sensitive or extremely important, then you may need your project sponsor to apply some pressure to free up the resources or even go outside the organization. Document these needs in a project charter (goals/objectives/problem statement) and a project plan (schedule, cost estimate, staffing plan) to be submitted for management approval.

Useful questions:

- Is this the most valuable problem to solve?
- Do methodologies already exist that are accepted for the problem at hand?
- What do you think is going on?
- What are some of the elemental relationships?
- What are the drivers that you think will move the needle on results?

Chapter 8 shows several approaches for developing and enhancing decision frames.

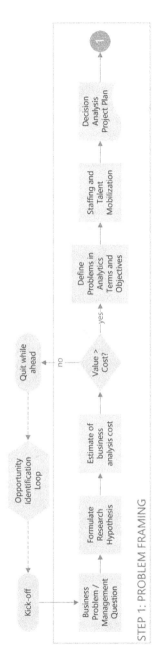

Figure 43: The problem-framing process.

Data collection and assumption research

Data collection is a process that can span many sources and many levels of abstraction. Transactional data is often available from the ERP or the business intelligence platform and can be a good starting point for your analysis. Of course, most models seek to account for exogenous variables and information, and this can be handled in the external data collection process using web scraping, information services, open data, APIs, and so on.

It is easy to drown in a sea of data, so navigating intelligently is vital. To this effect, we need to wonder whether we have the right kind of data or whether there are other sources that we have not considered that would enrich our analysis.

Conversely, a subjective analysis may be required because the event we are trying to predict has never happened or is a very low-frequency event and thus, no datasets exist. In these cases, the best strategy is to collect primary data, which can include conducting surveys, prediction markets, focus groups, interviews, or just bringing in a few subject matter experts to help you scope out ranges and assumptions for your model.

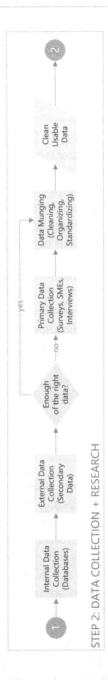

Figure 44: The data collection and research process.

Assumptions will fill the gaps where we don't have data and supplement the situations with little or no data. Sometimes, this is a good thing or you can end up putting your thumb on the scale, so you must do this with care and ideally when some sort of theory exists to support that assumption.

To highlight how assumptions can be self-serving, when I was at university, if the professor did not specify a detail in the case study or the problem, I would make the most accommodating assumption to my approach. This, of course, would make my life much easier, save many paragraphs of writing, and support my conclusion. This is *only* good practice for getting a good grade in university when there are no impacts to making these kinds of project-friendly assumptions.

Alternatively, some models have built-in assumptions that help when almost no data is available. Imagine we are trying to model the waiting lines at a bricks and mortar store. Because it is very expensive to stand around all day to note the arrival times of customers, we only have access to a small sample of the customer arrival times. In simulation or queueing theory, it is an accepted practice (going back to Erlang in 1909) to assume the arrivals are a Poisson process and use the sample to develop the Poisson distribution parameters that represent uncertainty (as an uncontrollable variable) in the analysis. Therefore, we need to remember that data availability will affect the definition of our assumptions.

In the absence of data, some large assumptions may need to be made, all while making more subtle assumptions through the acceptance or abstraction of certain details. Perhaps the data is a proxy for what is needed to get the job done right and this too is making an assumption. Tracking the assumptions is of critical importance for both auditing and communicating results. Check your assumptions before presenting your results and recommendations, and make sure the decision-maker is aware of them.

Getting the data can be challenging on its own, as it involves dealing with IT, governance, hidden spreadsheets, and so on. Once we have it, the data still needs to be cleaned, organized, categorized, and dimensioned. At this point, conduct exploratory analysis to further identify relationships, perform some quick visualizations, and look for trends that could figure into the predictive model. The translation of these elements happens in the model-building step.

Useful questions:

- What data are available?
- What are the assumptions you need to do this analysis?
- Do we have the right data?
- Do we have enough data?
- Does the data exist?
- Do we need to make many assumptions?

A historical example of when trend analysis leads to foolishness

A famous example of assumption making one look stupid is the famous photograph of Harry Truman after winning the 1948 US Presidential Election, holding a newspaper with the headline Dewey Defeats Truman. The newspaper had to go to press before knowing the election results, but assuming trends about voters would hold, they wrote a headline that Truman's adversary won—and was embarrassed.

Model building process

Model building is rarely, if ever, an exercise conducted in a vacuum. Models are always a response to some event that triggers the need to perform further analysis and often come from someone higher up. By understanding who the stakeholders (those who can affect the decision or be affected by it), project sponsors (the people who pay the bills and want results), and subject matter experts are, it is then possible to get a holistic perspective on the problem we are trying to solve. Having a perspective is important because *we can't boil the ocean*. We need to focus on what is important because we can't solve everything or answer every question with our model. If such a model was indeed possible, it would be so complex that it would be useless to us mere mortals!

Sometimes, the number of stakeholders is quite large, and we may need to limit our consideration to those who are influential in the decision. At the end of the interviewing process, we should write an opportunity statement and prototype a simplified version of the model to test the applicability and further validate with our stakeholders that we have correctly captured both the business logic and the decision they are trying to make.

Once we have built our model, we should do a quality check for *model verification, validity, and accreditation. Verification* is ensuring the model works correctly. Do the formulas produce the correct answers? We may use our model to explore ranges of inputs that will provide unknown answers. But if we have cases where specific sets of inputs should generate specific outputs, we should use them to test our model. If this is not possible, say with a model of a system that is only a concept, vary one input at a time to see the behavior of the output. While the output values may be unknown, their behavior should make sense. For example, if we double the input amount of cost, our profit probably should not increase. If it does, investigate because that is completely counter-intuitive. That said, sometimes the model is right, and the unexpected behavior becomes a valuable insight. But be sure that valuable insight is real and not an artifact of a flaw in the model.

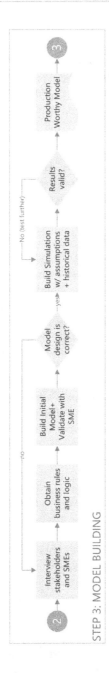

Figure 45: The model-building process.

Validation is ensuring the model is appropriate for its intended use. If we built a bespoke model to answer a question, this is not likely an issue as we built the model for that specific purpose. It becomes important when an existing model is used to answer a new question. That new question might be quite different or a slight shift in interest from the initial question. Often, these shifts are either of scope or scale. For an example of *scope*, imagine we have developed a sales model for consumer product x. We are asked to use it to represent consumer product y. The model might work well if the drivers of sales for both products are the same and if the sales are independent of each other. But if one product cannibalizes the sales of the other, then we will likely have problems. For an example of *scale*, suppose we have built a model to represent a project where a chemical plant will be built. We include capital expenses in evaluating various high-level designs, but then someone asks a more detailed question such as, "How much Schedule 40 8-inch steel pipe will we need? Maybe we can estimate the total steel cost range, but the model (estimate) can't provide that information without making a lot of assumptions.

Accreditation is the idea of certifying a model for a particular use. Reusing models appropriately provides a lot of efficiency. For example, if our company starts operating in a new country, it may be valuable to craft the tax calculation portion of our model so it can easily be used by other models. Accredited models are normally locked so the user can't accidentally damage the model but only change the input data. This avoids having to repeat verification each time we use the model. Documentation for accredited models is especially important.

If the model is for one-time use, then deployment is simple—it's only deployed to you for further analysis! However, if it is to be available for decision support in the organization, we need to carefully consider the user interface, documentation and training needed for successful use.

Collaboration with the various stakeholders is an important and ongoing task that will dictate whether we successfully build a great model and, more importantly, get it used. When building a predictive model, it will be necessary to go back to the stakeholders and subject matter experts to calibrate both the inputs and correlations among them—otherwise, garbage in, garbage out. We also need to consider how we will manage the model over its lifecycle. Periodically, we will need to assess whether the validation holds and potentially recalibrate the model because of changes in the environment.

Lastly, to make sure everything is running smoothly and as intended, we need to retest and validate the results. Just like in data science, we split our data into a holdout and a training set. This will allow for back-testing the model and seeing whether it would have predicted the holdout data correctly. Of course, given these changes, if the fundamentals of the problem have shifted, then there is no guarantee that a properly fitting model on historical data would perform well.

Analysis and optimization

The main deliverable of the analysis and optimization process is a decision strategy recommendation report with multiple viable options.

To deliver a robust recommendation, some of the most critical tasks for the decision superhero are identifying and ranking the best courses of action (COAs) for a decision-maker's problem. Ranking can be a qualitative or quantitative measure, but in our case, we will focus on the quantitative dimension. Furthermore, we feel there is a significant role for computer-aided assistance, that is, the use of optimization algorithms, to help further identify valid options.

The starting point of any good analysis is to have an **anchor point** (known as a base case). *Your base case consists of the results your model produces with your initial configuration of variables under the decision-maker's control (courses of action) and best guesses for those that are not (states of nature).* A base case result acts as your initial hypothesis for the results. You can find it in a variety of sources, including contracts, corporate documents/policies, subject matter experts, or in the public domain.

Once all base case assumptions and input variables are correctly defined, when using probabilistic methods, run a simulation to get a base case distribution of the output. The goal is to compare risk-based probability distributions versus single-point estimates to make a decision.

Are there better courses of action out there? Most probably. You can find these using brute-force[16] simulation/optimization methods like stochastic optimization. Stochastic optimization is a technique where you can *optimize the output distribution* instead of a single point like you would using conventional linear programming methods. This means that you can apply it to almost any problem, including non-linear ones, without issue.

---

[16] Brute force refers to Monte Carlo optimization which involves systematically testing a large number of possible solutions using random sampling to explore the solution space to find the best one.

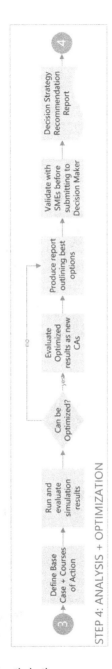

Figure 46: Model analysis and optimization process.

The basic process for stochastic optimization is to select a series of controllable variables for the model, run a complete simulation, and document the output distribution or the performance metric of interest (for example, Median, 90th percentile, Standard Deviation, and so on) in that COA's probability distribution. The concept of optimization is applied when you filter the COAs (combinations of decision variables) based on their results and constraints to select the best options/configurations.

If there are one or more better courses of action that are identified by running stochastic optimization, then we should include those in our analysis before submitting the results for approval. **Of course, as always, we must validate the results with subject matter experts and some of the key stakeholders to make sure that we are indeed considering the best options.** Once we have established our best course of action, it is time to write up a report with the best options and a recommendation for moving forward (or not).

Production models for ongoing analysis and model lifecycle management

Production models are models that have been debugged and scaled to the size of the business. These models have the characteristic of being run multiple times, such as rolling forecasts.

If it is recognized that the model is to be run periodically, then careful consideration about moving it to code is required because that can translate into some sort of competitive advantage based on velocity (because as we keep saying, speed in decision-making is a competitive advantage).

At this point, the model should already exist in some form, such as a spreadsheet, a computer workflow, or even a drawing on a piece of paper. Just like in any other computer project, the first step is to start your business analysis process to translate your model into code that reflects specifically what the business is asking for.

Business analysis using 3-tier design principles

As part of the business analysis, the next step is to take these specifications and plan a practical project to deliver the most pressing components to the business. The best approach to getting more funding is to use an Agile approach. That is, break out your project into smaller components that can deliver value immediately and justify further investment in your initiative.

The three-tier design, a staple in software architecture, serves as an effective blueprint for organizing systems across three distinct layers: *the presentation tier, the business logic tier, and the data access tier.* This structured approach not only facilitates system development and maintenance, but also enhances scalability and security across various applications.

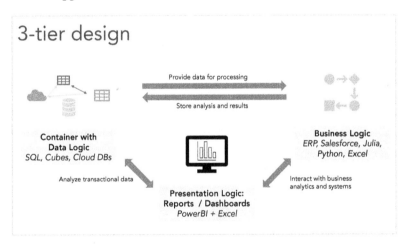

Figure 47b: 3-tier design applied to analytics and decision science.

Figure 47: Production models and lifecycle management.

The presentation tier forms the uppermost layer of the architecture, where user interaction with the system occurs. This layer is crucial for shaping user experience through the interface, allowing users to input data and receive results. It operates independently of the deeper layers, enabling designers to focus solely on user interface issues without delving into the underlying business logic or data management concerns. Beneath the presentation layer lies the business logic tier, which processes user inputs, applies business rules, and makes decisions. The data access tier is at the base of this architecture, responsible for managing all interactions with the data source.

You may not realize that Excel supports the development of models that use one, two, or all three tiers. The worksheets that serve as the user interface represent the presentation tier. You can design these sheets as dashboards or applications that display data, charts, and input forms that users interact with. Excel's capability to embed formulas, functions, and macros (VBA) serves as the business logic tier, processing the data entered in the presentation layer and performing calculations, data analysis, and decision-making tasks. Although Excel is not a database, worksheets dedicated solely to data storage and supported by PowerQuery can simulate the data access tier to perform queries and data tasks. When building out Excel models, many analysts start with the data tier by putting all the data into SQL and connecting to the data. This results in models that can handle way more data and process it in error-free ways, which is not always a guarantee in Excel.

Applying the three-tier design to complex decision models within large enterprises or data-intensive scenarios allows for streamlined delegation of tasks across various specialized teams:

- *Presentation Tier for Interface Designers*: Assign a team specifically to develop the user interface, focusing on user experience and interaction without being bogged down by the underlying business logic. This team ensures that the decision-making process is accessible and user-friendly.

- *Business Logic Tier for Analysts and Subject Matter Experts*:

This group focuses on the core decision-making processes. They implement the models and algorithms that drive
business decisions, translating complex rules and logic into
programmable steps. This tier serves as the bridge between
raw data and user-facing outputs.

- *Data Access Tier for Database Administrators and Data
  Engineers*: Specialists in data handling focus on managing
  how data is stored, accessed, and optimized. They ensure
  data integrity and security, providing a reliable foundation
  for the business logic tier.

This framework supports parallel development, where each tier can
be developed independently by teams focused on specific aspects of
the application. For example, while the user interface team optimizes
the layout, the business logic team can develop decision algorithms
independently, and the data team can simultaneously enhance data
storage and retrieval systems.

Moreover, this separation of concerns facilitates easier updates and
scalability. Implement changes or enhancements within one tier
without disrupting the others. For instance, updating business rules
in the business logic tier does not necessitate changes in the user
interface or the data access protocols. Additionally, focusing security
measures on the data access tier helps protect sensitive information
more effectively without burdening the other layers with unnecessary security complexities.

In summary, the three-tier architecture not only enables organizations to manage the work associated with developing and managing
complex decision-making models more effectively, but also enhances the overall agility, scalability, and security of their systems.

After coding the model, you need to test it. A barrage of tests should
be applied both to the components, to ensure that the functions are
working consistently, and to back testing, to ensure that the predictions work under uncertain conditions.

- Have a gold standard to test results
- Have the model tested under extreme conditions using simulation and sensitivity analysis
- Use other recognized methods to generate comparison results that tell you that the model is consistent.
- Have fellow decision superheroes review the work

In life, just as in modeling, it's all downhill after you are born. Every model has an unknown lifespan for prediction accuracy because of the ever-changing dynamics in the environment. Sometimes, addressing this can be a simple question of updating the inputs using an API, while other times, you may need to conduct some form of fundamental analysis with subject matter experts because the data does not exist in sufficient quantities to be able to use numeric methods. This ongoing updating and calibrating help mitigate what is known as *model deprecation. Deprecation is a fancy word that means you lose performance over time due to changes in underlying logic or inputs that are not being captured or updated.*

To stay accurate, a model is an iterative process unto itself. Every time a model's assumptions are updated, it is important to document the modifications and the accuracy that these modifications brought to the model for future analysis. If the model is running within tolerance, then it needs to be put back into production and re-evaluated at some pre-defined point in time, or if the performance metrics of the model fall off a cliff before then. For example, when COVID-19 hit, given that all air traffic suddenly stopped and then the volume was considerably restricted, the airline pricing and scheduling models went nuts and could not correctly predict what was needed due to such a drastic change in the underlying data. Forecasting with market data was similarly impacted when the 2008 crisis happened.

Of course, every time you iterate through a model, you must ask yourself whether you need to fix it, replace it with something better, or stop maintaining it because the business need has gone away. It is amazing how many models and reports get maintained long after they have stopped being used, thus creating lots of confusion and messiness for those who need to use them.

Driving the decision science roadmap buy-in with quick wins

The most compelling reason for getting early adoption and buy-in with quick wins is that it ensures a steady flow of budget and projects for you and your decision superhero colleagues. Beyond the notion of self-preservation, the *Institute for Operations Research and the Management Sciences (INFORMS)* recommends simultaneously keeping in mind both short-term and long-term objectives/benefits when planning to bolster the adoption of quantitative approaches. Firstly, addressing short-term actions, which build momentum, and secondly, configuring a long-term strategy to support resource planning and avoid pitfalls. Short-term actions are often referred to as *quick wins*, as the intent is to demonstrate value rapidly. This provides a quick answer and allows more wins in a period. *If you have six months to demonstrate value, it is better to have three smaller successes rather than a single larger case.* There are several good reasons for this, including the following:

Multiple successes increase confidence in the results

- The probability that all three projects fail is low (naively estimated as 50% x 50% x 50% = 12.5% chance of all three projects failing)

- Having successes in different parts of the organization demonstrates the broad applicability of the solution

- It encourages additional stakeholder managers to be supportive

Staffing and resourcing for quick wins

When considering what projects to establish as quick wins, it is important to consider the required resources. In the short term, you need to have the resources available to support your quick-win projects. These resources include people, tools, data, and information technology. In the long term, your resourcing may look quite different.

Depending on the size of the project, you might be the sole analyst or the project may be big enough to require additional people to do the work. If the work is new to your organization, even if your level of competency is sufficient, finding others with the right skill sets can be a challenge. Thus, drawing on external expertise to help with these projects may be necessary. Eventually, the goal may be to develop sufficient internal expertise for the work, but having experienced coaches will accelerate this process and be helpful during the quick-win phase.

Experienced *external coaches* can help you deliver the project and they can also provide feedback about the project versus experiences with other organizations. Though it is often taken for granted, making sure that the resources you intend to use are truly available can make life much easier. You might receive an employee's name with good intentions, but if they still come with obligations related to their regular role, they may not be able to contribute as needed for the project's success. Another cynical consideration is that the least talented or productive people are usually the most available. Carefully manage who gets sent over from the bench. Otherwise, this could triple the work or doom the project entirely.

Tools and methods for quick wins

If this is a new capability for the organization, the tools needed may not be available. Some tools can be readily developed using the techniques in this book, but time should be allocated for tool development, as well as doing the project itself. If specialized software is needed, make the purchase budget available upfront. Once the project starts, the need for different tools may become apparent, far different from what was anticipated. Thus, a contingency factor in the budget may become important.

Picking winners and not betting it all on one horse

The most important aspect of doing quick-win projects is likely the selection of the projects themselves. Ideally, the projects will have supportive managers and the support of the users of the study. The study results might be a recommendation for the decision, but they might also be a decision support tool for future use. The user community can be a significant stakeholder group. *If the user group is not supportive, such as when the modeling approach differs from the group's experience set, the group may be hostile.* The analysts conducting the study must have enough domain knowledge to understand the nature of the problem and communicate well with users.

Managing for success

Given the resources and time allowed, the project should have a reasonable chance of success. The project should be visible to members of the organization, and the projects should be impactful so that success is readily valued. It is helpful to have clear benchmarks or goals against which to judge success. Alignment of the project and the resources can be a bit of a chicken-and-egg problem and may require iteration.

In general, recognize that these early successes are necessary to establish a foundation to build improved organizational DQ. Eventually, obtain support from senior levels of the organization. Middle management often conducts the day-to-day business in the most profitable way, and may not be heavily exposed to the organization's strategic thinking. As a result, they often are the most reluctant to adopt new procedures. This is known as dealing with the *frozen middle* in organizational change management.

Conduct the short-term quick wins with an eye toward an anticipated long-term strategy for implementing broader organizational
changes. However, as you conduct initial projects, you and the other
organization members will learn and form opinions about their implementation. The suggestion is to always be ready to recommend
deploying a permanent capability while continually revising the
potential recommendation as implementation progresses and you
learn more.

Key takeaways

- A Decision process can be optimized to be faster or better
  and hopefully both.

- Exploring data is to inform and prediction is to make
  decisions.

- Have a master process for organizing to maximum effect
  decision science activities within the firm.

- The three-tier design can streamline the implementation
  of production models by breaking out the work in terms of
  visualizing the results, developing the decision model, and
  managing the data used to do the analysis.

- Delivering results to the business is key to getting more
  money for decision science.

- Work on projects where the project manager and stakeholders understand the value you bring to the table.

EIGHT

Decision Design

Designing a decision involves defining the criteria and processes for making decisions and structuring the environment so you can make decisions that optimize outcomes. Decision design can encompass various aspects, from the simple layout of choices to complex, strategic decision-making problems.

Starting with the core definition of a decision, we will build on this by exploring some of the best practices, major considerations and elements of a decision, and its supporting analysis. Then, we will delve into framing in greater detail, and go through communication methods, questions, and workflows that allow a firm to get to a better, clearer decision faster.

Effective decision design is critical because it helps ensure that decisions occur systematically, transparently, and informedly, which can significantly impact the success and efficiency of organizational operations. It is especially relevant in business strategy, policy formulation, resource allocation, and any scenario where choices determine significant outcomes.

The elemental definition of a decision

A decision consists of a situation, a constraint, and a preference for outcome. A decision is an irreversible allocation of resources to an outcome — meaning if you do one thing, you are not doing something else. A decision is atomic but can be part of a greater set or portfolio of decisions.

Decisions vary in complexity; some necessitate extensive mathematical analysis, others need only basic calculations, and some require none at all. Choosing a favorite color is purely a matter of *preference*, devoid of any mathematical consideration. Conversely, selecting a car or deciding on a neighborhood involves more *quantitative assessments*, such as fuel efficiency, payment schemes, and resale value—each of which may be subject to some constraint, be it real or perceived. However, with subjective preferences, you can simply like the yellow one that goes vroom-vroom!

Preferences and constraints

A preference is simply what you like. If you like strawberries but hate bananas, it stands to reason you will pick a strawberry daiquiri over a banana daiquiri every time. In reality, sight, taste, sound, smell, and touch can highly influence your preference for situations and objects. For example, if you don't like loud noises, you might prefer going to a library rather than a heavy metal concert. That being said, no one is saying you can't like both.

If you give such importance to a preference that it eclipses all other considerations, then it becomes a constraint onto itself.

A constraint is a limit placed on a variable. These limits can set a lower bound, an upper bound, or both. For example, a car must:

- be less than $35k
- It must have a fuel efficiency greater than 35 miles per gallon (or 6.72 liters/100km)
- It must be all-wheel drive.
- I want it yellow
- Etc.

When you look at these as hard constraints, you will immediately discard any option that does not meet one of these considerations. If this limits your options unduly, you need to soften your constraints to increase the number of available options. For this reason, you might accept only a partial set of constraints. For example, the car is less than $35,000, has all-wheel drive, and acceptable fuel efficiency, but is only available in gray. If your color is a constraint, you will decline the car, but if color is a preference, you may consider that a good trade-off even if the color is not perfect. That decision would have satisfied all the constraints. If there were other alternatives, then decision modeling would be used to make an "optimal" choice between available alternatives/courses of action.

Setting the value context of a decision

As the old saying goes, context matters. Context can define the selection of inputs, the way to apply the model, and how to define success. The relative value of the alternatives can take many forms, but in most cases, it tends to be monetary. In a public sector context, evaluate the alternatives against other non-monetary but concrete metrics, such as wage levels, health outcomes, homeless levels, etc. On the other hand, there are problems where the measurement takes on a more abstract and harder-to-define subjective and relative criteria. To make this point concisely, the authors of *Algorithms We Live By* suggest that this can even hold true in romantic pursuits. Gold digging, that is, the act of seeking the right romantic partner based on money, is a very efficient selection criteria because the measure is absolute and mathematical: money. The candidate with the most wealth wins! Alternatively, a selection process where one may place a higher value on other desirable abstract qualities that are far harder to measure, such as trust or empathy, makes selecting the optimal result more arduous. Add to this that personal experiences influence the decision criteria, and you quickly realize that the choice is no longer so clear-cut. The primary reason is that the abstract nature of the measurement problem makes deciding on the right candidate far more subjective, especially because of the qualitative versus quantitative nature of the assessment.

When Alfred P. Sloan wrote his book, *My Years at General Motors*, in 1963, he outlined many decision-making processes pioneered in the early 20th century to allocate resources. They created the DuPont model, ROA (return on asset), ROE (return on equity) valuation methods, and project and product portfolio optimization.[17] Sloan emphasized that the cleanest and easiest relative measure among options was money because it was the only real common denominator across the organization. He extended this thinking to equally include ratios and performance indicators. This approach is now commonplace almost 100 years later but should not be confused as the only one.

As mentioned above, in contrast to strict monetary analysis, another widely held economic perspective for evaluating decisions is one you find in the public sector. The goal of the public sector (in most liberal democracies) is to maximize social welfare while minimizing inequalities for the broader good. Since you no longer have a universal measuring stick, this implies framing the problem differently, focusing on a different methodology for measuring performance and valuing outcomes. For example, where healthcare is a public service, performance is not evaluated on a monetary basis but rather on the outcomes, such as the number of visits to the hospital, patient return visits, days where beds were occupied, and so on. The basic performance management relationship between metrics and outcomes re-

---

[17] We cover the math and how to build portfolios using these methods using Excel and Julia in Book 3.

mains unchanged, but the units change. With apologies to Marshall McLuhan, "The metrics are the message!" How you define a decision's performance is truly a question of context.

Who's making the decision?

Knowing who makes a decision and how that person makes a decision need to be known upfront by all involved to avoid any issues or conflicts between decision-makers who may feel they each own the decision. It also sets the tone for discussions with stakeholders, who may influence or be impacted by a given decision. There are four basic ways to make a decision:

The Consensus approach involves the entire group in the decision-making process, aiming to reach a decision everyone can agree to or at least support. It encourages open discussion and considering different opinions and perspectives to reach a collective agreement. The benefit of consensus is that it fosters team unity and commitment to the decision outcome. However, it can be time-consuming and may lead to compromises that dilute the effectiveness of the decision.

Fiat decisions are made by a single authority or leader without necessitating agreement or input from the rest of the group. It can be effective when quick, decisive action is needed or when the decision-maker has a clear vision or expertise that guides the decision. The advantage is speed and clarity, but it can lead to issues with buy-in and acceptance from the rest of the organization if not managed carefully.

In *structured team decision-making*, you follow a specific process or methodology to guide the team through making a decision. This could involve defining the problem, generating alternatives, evaluating those alternatives, and then making a decision. The structured approach ensures a thorough examination of the decision and its potential impacts but requires a well-organized process and can be more time-consuming than other methods.

Dialectical Inquiry divides the group into two subgroups to develop and then debate opposing assumptions or recommendations. It uncovers the best solution by rigorously examining conflicting views. This approach is beneficial for thoroughly exploring complex decisions but requires a mature group capable of handling conflict constructively.

Each of these methods has its context where it can be particularly effective, depending on the nature of the decision, the urgency, and the organization's culture.

Basic elements of a decision model

Let us decompose the main elements of any business decision used to build useful and easy-to-understand predictive models. In later chapters, we will further detail these elements from a practical modeling perspective, but let's get the concepts straight for now.

The objective of the decision

The objective function is the output of the system. The objective of the decision-maker is to choose a course of action that leads to the optimum value of the objective function. They may seek to maximize sales, minimize costs, or maximize profits. The word *optimize* often means maximize what we want more of or minimize what we wish to reduce or eliminate.

Most, if not all, business decisions fall into one of two categories: minimize (or avoid) disaster or maximize gain. If your goal is to improve sales, profits, market share, and margins as well as a host of other key performance indicators, then it is likely that you are seeking to maximize the results by picking the best configuration of the variables under control. Conversely, the analyst or decision-maker may seek to minimize defects, service incidents, process breakdowns, machine breakdowns, and so on. Again, the objective and context of the problem will dictate which direction the optimization will go in.

Through practice, you will notice that correctly identifying the objective function is one of the most difficult aspects of decision-making. If the objective is not stated or obvious, understanding what to measure as the key indicator of a decision's performance is usually quite definitive. Some may refer to this as the objective function of the model. By carefully analyzing the results of the objective function for different states of nature, you can have a basis/justification for making a decision.

Controllable input variables

Every decision problem, once its objective is clearly defined, involves one or more variables (or inputs), which are subject to the control of the decision-maker/scientist. The most foundational aspect of decision-making is the selection of the appropriate parameters (values or settings) for these controllable inputs. Classic examples of controllable variables in business can include price, quantity ordered, number of salespeople, marketing budgets, transportation routing, cities for plant locations, and anything else within the decision-maker's scope of control.

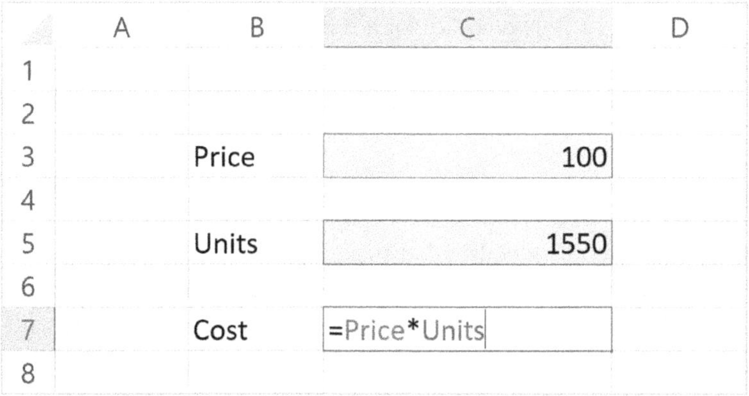

	A	B	C	D
1				
2				
3		Price	100	
4				
5		Units	1550	
6				
7		Cost	=Price*Units	
8				

Figure 48: Calculating cost using controllable variables (Excel).

We may imagine the decision-maker is sitting at a console with a series of one or more controls (labeled price, quantity ordered, and so
on) where they can plug in values and see the impact on the output
of interest. A contemporary version of this would be to set up an Excel model with inputs you can set yourself. Of course, the objective
for the decision-maker is to set parameters they find most optimal
for their purposes. Any distinctly identifiable configuration of these
controllable variables is called a *course of action*, and any change in
any one of these controllable variables produces a different course
of action.

Uncontrollable (random) input variables

One or more input variables in the system are not subject to the
decision-maker's control but affect the output under measure and,
therefore, require identification and, if possible, measurement. These
variables are mostly modeled as probability distributions[18] (Normal,
Poisson, Exponential, Weibull, and so on) and can be unitary or de-
composed into many sub-variables over time. They may be global
to all models or local to a specific application or area. Uncontrollable variables might be inflation, escalation rates, competitor pricing,
supplier pricing, customer income, level of demand, competitive ad-
vertising, meteorological variables, government actions, and so on.
We call any distinctly identifiable configuration of these uncontrollable variables a *state of nature* or a *background*.

Courses of action

A *course of action* is a possible option or alternative on which you
want to predict the likelihood of success. To build a model, the framing process relating to courses of action must assume that the decision scientist (and decision-maker) knows all the relevant alternative courses of action.

---

[18] https://en.wikipedia.org/wiki/List_of_probability_distributions

History is full of examples demonstrating that this may often not happen. In almost all areas of human pursuit (business, government, defense, and so on), one of the most important functions of a decision scientist is to seek out and identify all the relevant available courses of action. This requires an awareness of the system, processes, and institutions involved in the decision, as well as a good dose of curiosity and creativity. You can also use mathematical optimization techniques to further identify combinations or courses of action not initially considered but may deliver better results under uncertain conditions. After identifying these alternative courses of action, you need to select among them. This is where we switch to quantitative decision science methods to stress test the courses of action and mathematically recommend a superior course of action (if any exists).

States of nature

A *state of nature* defines a set of modeling assumptions (deterministic or probabilistic) under which the model will be run and tested. States of nature are often defined using uncontrollable variables and similarly, common quantitative decision-making assumptions include interest rates, FOREX rates, market demand, weather conditions, amount of oil in the ground, proportion of people willing to buy online, and so on. Because not all states of nature are known, using Monte-Carlo simulation becomes necessary to manage the analysis of the uncertain variables on each course of action.

Mathematical and/or visual relationships between inputs and outputs

The various possible configurations and combinations of the controllable and uncontrollable input variables (states of nature) interact within the system and produce a corresponding output, the objective function.

It is this functional relationship that the decision superhero attempts to discover and describe in mathematical terms. Considerable effort and a good deal of research go into identifying and describing mathematically the relationship between various variables and or entities. There are many methodologies for doing so, which include process mapping and redesign, systems design, network analysis, influence diagrams, decision trees, and so on. By thoroughly understanding the problem under study, a domain expert should already be aware of some of the more fundamental relationships before trying to organize the information into a model. We shall cover models and modeling in greater depth in Chapter 5.

Decision framing like a champ

"It is not important what you know. What is important is what you understand!" - Professor Julius Sumner Miller

In Chapter 7, we explored a structured organizational process for framing with a few pieces of helpful advice. In this section, we will look at actual framing approaches and best practices you can use to organize your decision problems.

Successful decision projects rely on a host of soft skills and qualitative methods that define the research question or the problem to solve in clear and concise terms while focusing on the importance of communication throughout the process. Frequently, qualitative observations in nature, usually some sort of observation, **may trigger the framing of a decision**. For example, a person is a drop-out, a product is defective, someone prefers to shop online instead of downtown, or a contractor prefers a supplier's products over another.

A disciplined framing process is the basis for most successful projects and should be considered the starting point of any new decision. Framing a decision defines the scope of the course of action you are trying to elect and why. A frame, just like any picture hanging on your wall, is the boundary between what is in the picture and outside the picture. If you do not set a scope (frame) for your decision, there are no theoretical constraints or trade-offs to use in your model or code to select an optimal course of action.

> *You can't get there from here if you don't know*
> *where here is, or there for that matter!*

The easiest way to tackle the framing process is to look at it as a series of questions that you need to answer in a certain order:

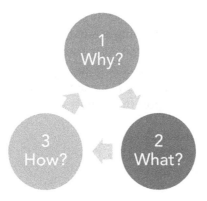

Figure 49: The hierarchy of framing questions.

Given that decision science is an iterative approach, you may cycle through these questions multiple times before settling on the scope, data, and methods needed to make a decision. From a modeling and programming perspective, understanding these items lets you know where to look for data, the APIs to use, the logic and the formulas used to calculate the metrics for your decision, and what your final analysis should (tentatively) look like.

Why?: *Why are we doing this and not something else?* This should be the starting point for any discussion. Knowing why we are doing something sets the context and the urgency for the model we are developing. Some useful and necessary techniques for getting to Why include:

- Brainstorming
- Research
- Whiteboarding
- Work-out process (originally designed by Dave Ulrich and Ron Ashkenas while at GE)

What?: Figuring out what invariably starts with why! Once you know why, then you can start looking at what to do. What is the solution/ model we are trying to develop? What should it do? What is available on the market? The process of defining what is akin to defining the options under evaluation. Any of the proposed solutions included for analysis need to support the realization of Why:

- Supplier meetings
- Market analysis
- Competitive actions
- Expert elicitation

How?: Once you have identified the options to analyze and contrast, you need to define the frame in specific terms to allow for both the creation of a model that reflects some sort of reality as well as one you can put into a practical form that can be shared, discussed, validated, and agreed upon.

Each audience has its role to play in decision making - be it strategic, tactical or operational (respectively known as *why, what* and *how*). The sponsor is the root of all success (or failure) because they are responsible for obtaining resources and corporate attention, we put it at the center of the stakeholder development process (Figure 50)

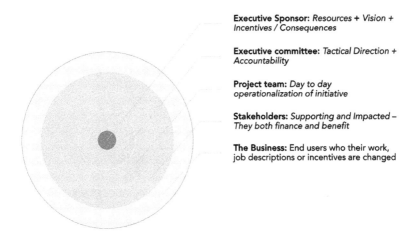

Executive Sponsor: *Resources + Vision + Incentives / Consequences*

Executive committee: *Tactical Direction + Accountability*

Project team: *Day to day operationalization of initiative*

Stakeholders: *Supporting and Impacted – They both finance and benefit*

The Business: End users who their work, job descriptions or incentives are changed

Figure 50: Framing decisions with your audience in mind

Each audience has it's particularities. To make things easier, we have summarized what drives each audience to help you be more effective tailoring your message and getting your point across effectively.

TABLE 9A: ARE YOU ANSWERING THE RIGHT QUESTION? UNDERSTAND YOUR AUDIENCE.

Audience	Focus Question	Drivers
Executive Sponsor	**WHY**	Achieving a stated strategic objective
		Regulation
		Personal success (promotion, recognition…)
Executive Committee	**WHAT**	Measurable results
		Coordination among functions
		Ensure success via allocation

Audience	Focus Question	Drivers
Project Team	**HOW**	On-time delivery
		Staying within scope
		Budgets
		Minimizing disruption
Stakeholders	**WHAT**	Sharing in the initiative's benefits
		Personal success (promotion, recognition…)
		Provide direction on things that impact them
The Business	**WHAT + HOW**	Mitigate loss of personal status
		Feel comfortable in new environment
		Understand how the change is personally beneficial

Better questions and communication

As a decision superhero, *the ability to ask the right questions is imperative, along with the ability to capture the information that is important to you when an answer is provided.* Sometimes, knowing where to start is the toughest aspect of solving a problem. Usually a sound approach is breaking out complex problems into smaller more manageable components; as the old adage goes, "Do you know how to eat an elephant? One bite at a time!"

The 3Rs

The 3Rs refer to a communication technique designed to ensure that the intended message is making it across successfully in a conversation. Sometimes, the language or common examples might differ due to language ability, domain knowledge, or background differences. **Always remember that successful communication is the responsibility of the emitter.** Here are three techniques to ensure proper understanding:

- *Repeat*: Repeating works both ways. You can ask the other person to repeat back to you what they got from your instructions, or you can loop the explanation one more time.

- *Rephrase*: Pick different words to express the same idea. Sometimes jargon gets in the way.

- *Reframe*: Switch or propose a metaphor. Imagination can be very powerful.

In figure 51, we show how to use rephrasing and reframing together to improve the understanding of a decision frame when interacting with experts and stakeholders.

Asking Framing Questions Using **the 3Rs**

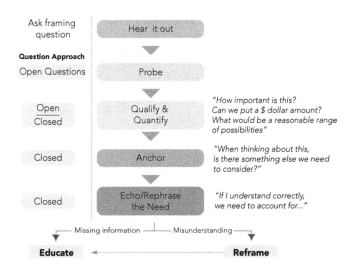

Figure 51: Framing process using the 3Rs.

When the 3Rs combine with a powerful question framework such as the 5Ws, you end up with 15WRs... ok, just kidding. But in all seriousness, they work hand in hand and support a better conversation.

The 5Ws

The 5Ws (Who, What, Why, Where, When) offer a timeless and effective framework for problem framing, widely recognized across disciplines ranging from journalism to business analysis and beyond. This approach systematically explores the various dimensions of a problem or opportunity, addressing essential aspects to understand its scope and paving the way for solutions. Applicable across fields such as finance, project management, IT (including Infrastructure, Analytics, Big Data, and Strategy), public administration, and manufacturing, the 5W's help dissect problems to their core components.

Additionally, the utility of the 5Ws extends beyond merely gathering information; it plays a crucial role in disseminating and eliciting information. This is crucial when engaged in a business transformation, especially *in the early information and communication stages* of the adoption curve (see Chapter 3). Depending on the context, you can employ this framework to convey a comprehensive understanding of a subject to an audience (pushing information) or formulate incisive questions that uncover deeper insights (pulling information). This duality enhances its versatility, making the 5Ws an invaluable tool in both communication and inquiry processes.

TABLE 9B: THE 5WS FRAMING QUESTIONS FOR ELICITING AND
DISSEMINATING INFORMATION.

	Analyst Solution Design Questions (Pull)	Getting your audience on-board (Answers)
WHO	This question obviously covers the subject of your analysis. The people impacted, benefiting, or being hurt by your project. · Who does it? · Who is doing it? · Who should be doing it? · Who cares the most?	Who is the focus of your communication or receiving the information? Do they have particular sensibilities that you need to be aware of (political, religious, etc.)
WHAT	What are we building/doing? What is involved? What resources do I need? What should be done? What else can be done.	This is WHAT we are doing What will change? What you need to learn? What you need to know?
WHY	Why do I care? Why is this important? Why should I do this and not that? Why do it there? Why do it then?	WIFM: What's in it for me How people benefit around you Cost/Time/Quality improvement to stakeholders

	Analyst Solution De-sign Questions (Pull)	Getting your audience on-board (Answers)
WHEN	Are market conditions dictating when to act?	Explain the best timing and how that makes a difference financially or operationally.
	When will we have all the resources to execute?	Key milestones and what we will have accomplished
	When should we execute?	
WHERE	Where to do it?	Locations for convenience
	Where is it done/processed?	Online vs. Physical World.
	Where else can it be done?	Points of interaction and data generation

Utilize these straightforward questions to deepen your impact. When your analysis resonates and aligns with your audience, presenting outcomes in ways they understand and value enhances your contribution significantly.

Framing a decision using a payoff matrix

Imagine you run a food truck in New York City and want to optimize your profit (*Why*). You need to figure out what you should bring with you on any given day to meet your customers' needs. On a rainy day, history has proven that we sell more coffee and donuts; on a sunny day, history shows that we sell more ice cream and tacos. The predictive decision-making model (*What*) we seek to build is to figure out what we should bring, on whatever day, to ensure the service levels we have accustomed our customers to, most notably having what they want in the truck when they ask for it. Organizing your courses of action and states of nature is the How when it comes to framing.

The following graphic organizes the various outcomes we seek to analyze for our food truck. The outcomes in red (-) are those we seek to avoid, and those in black (+) are those we seek to maximize.

TABLE 10: FRAMING DECISIONS INTO OUTCOMES.

Course of Action	States of Nature	
	Sun	Rain
Ice Cream + Tacos	Win/Profit	(Loss)
Coffee + Donuts	(Loss)	Win / Profit

Analyzing or interpreting the results of a predictive model allows us to make sense of what the model says to us and how to act under various circumstances and conditions. Once framed, this is where most quantitative methods come in and when you start developing spreadsheets or coding.

Framing decisions in an organizational context

The 3-What Framing Approach uses a three-column table with the headings "What We Know," "What We Don't Know," and "*What Matters.*" It is a structured approach to decision-making that helps clarify the decision context by separating facts, uncertainties, and objectives.

What We Know: This column lists all the relevant facts, data, and information you currently have about the decision or problem at hand. It includes any research findings, historical data, proven assumptions, and other concrete information that can influence the decision. By filling out this column, you focus on grounding your decision in reality and ensuring that you consider all known factors.

What We Don't Know: Here, you identify gaps in your knowledge, uncertainties, risks, and any assumptions that you might need to make in the absence of hard data. This column is crucial for risk management, as it helps you understand what factors could influence the outcome of your decision but are currently outside your knowledge base. It encourages a proactive approach to seeking more information, preparing for contingencies, or deciding how much risk is acceptable in decision-making.

What Matters: This column identifies the important objectives, values, and priorities in making the decision. It helps clarify what outcomes you seek, what trade-offs might be acceptable, and what criteria you will use to evaluate the decision options. This section ensures that the decision aligns with the broader goals, values, and mission of the organization or individual making the decision.

TABLE 11: THREE WHAT FRAMING APPROACH.

What we know	What we don't know	What matters

Populating a 3-*Whats table* is straightforward. Here are four simple steps:

1. *Compile Information*: Gather and list all relevant information under each column. Engage stakeholders in contributing to each section to ensure a comprehensive view.

2. *Analysis and Discussion*: Use the table as a basis for discussion among decision-makers and stakeholders. Analyze the information to identify key drivers, constraints, and the potential impact of unknowns.

3. *Identify Gaps and Actions*: Determine if additional research or information is needed to fill critical gaps in "What We Don't Know." Decide on actions to obtain this information or mitigate risks.

4. *Prioritize and Decide*: Based on "What Matters," prioritize the factors influencing the decision. Use this prioritization to guide the evaluation of alternatives and make a decision that aligns with your objectives and values.

This method not only brings clarity and structure to the decision-making process but also promotes transparency and inclusivity by making explicit what is known, what is not, and what criteria are being used to make the decision. It can be particularly useful in complex decisions, where the interplay of different types of information and values can significantly impact the outcome.

Good decisions keep the trade-offs in mind

Effective decision-making can be worse in organizations where multiple stakeholders complicate things. For example, there was an oil and gas project where a multi-disciplinary team argued over alternatives for the target of an appraisal well. Arguing over alternatives, rather than the trade-off of the value delivered, is a bad sign. Working with the team to clarify and frame the project's sub-objectives helped make the right alternative clear. *Though an entire team supported the idea of a successful project, they all had different interpretations of what that meant based on what they valued, their role, and their professional history.*

The situation is more complex when structuring processes and decision-making within business relationships. Some industries commonly form *joint ventures (JVs)* for specific purposes, such as entering new markets, launching new products, sharing risk, and so on. Formed by competing companies, these JVs help manage risks and provide other advantages. Conflict between the partners is often a key contributing factor when they fail. Conflict can arise in many places if certain things are not done to foster trust and collaborative decision-making. Furthermore, the mathematical models used to support decisions are crafted in parallel by the different stakeholders. Stakeholder predictions and decisions differ because they use different data sources, models, and methodologies. Just like in the old cowboy movies, this is like showing up to a Mexican stand-off but with business analysts instead of guns–a situation made even more difficult when no one is right and no one is wrong based on the simple fact that there is not one single source of truth.

It is only natural for different organizations to have varying decision criteria that sometimes reflect subtle differences in their objectives. To this effect, Peter Senge mentions in the book, *The Fifth Discipline,* that having a shared understanding allows for having a personal interpretation of what to do to achieve a shared objective. Accounting for such nuances leads to building better models that incorporate more of the dynamics of the decision being modeled. For example, a large company may most value alternatives that maximize investment efficiency to support their portfolio of investments, while a smaller partner in a JV may be interested in maximizing the net present value of the activity. Another partner may be most interested in cash flow in specific years because of their circumstances, while another may want to see their presence increase in a new market. All these objectives support making a profit but do it in ways specific to each partner's needs and performance requirements.

Conversely, this type of situation is far more common within a firm than *between firms.* Generally, the stakeholders are unhappy because they have to make concessions that they feel make their solution less than optimal. This is called *satisficing* (at the lower level) and will lead to a global state that will work better overall, even though that is not how people generally feel about it.

Working-out decisions

Originally created at GE, the Work-Out method[19] is a decision-making and problem-solving process developed at General Electric during the Jack Welch era that combines agile decision-making with a collaborative twist. The core purpose of decision science is to gain clarity for decision-making and maximize the participation of rel-

---

[19] (Ulrich, Ashkenas, & Kerr, 2002)

evant stakeholders and experts. The Work-Out process integrates these elements in a practical way. Here is an adapted version of the Work-Out process to get from opportunity to decision rapidly and collaboratively:

- *Initial Sessions*: These work sessions involve teams of internal experts close to the issues, along with customers, suppliers, or business partners, as appropriate, to analyze business issues and develop improvement recommendations.

- *Executive Decision-Making*: Before concluding the Work-Out session, relevant executive decision-makers must publicly judge all team recommendations on the spot. This ensures accelerated decision-making and solid leadership endorsement.

- *Agile Implementation*: Teams reconvene at 30, 60, and 90 days to review progress, with most work expected to be completed by the final checkpoint. This feedback loop makes sure we don't waste resources. If a project turns out to be less impactful than thought, we remove it from the portfolio at review.

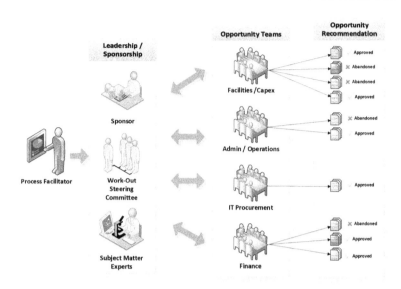

Figure 52: The Work-Out process .

The Work-Out process is unique for its ability to mobilize commitment and support across organizational boundaries by encouraging broad participation and supporting those brave enough to carry forward and execute their own best ideas.

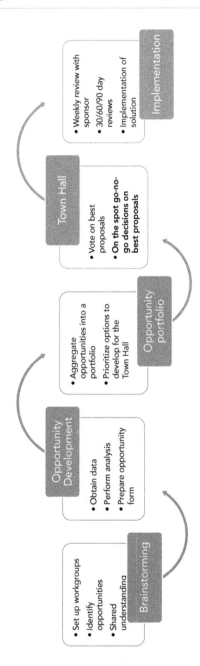

Figure 53: Work-Out process for decision-making.

Key takeaways

- Decisions are irreversible allocations of resources

- Decisions consist of a preferred outcome, preferences, and constraints. You need to think about and prioritize these.

- We often base decisions on considering one or more model metrics, and setting the value context or how to interpret the metric is important.

- Decision models consist of things we control and some that we do not. Understanding these and the relationships that connect them is critical.

- 3R and 5Ws help with the framing process to clarify what people mean and how they perceive the decision's structure and design.

- The Work-Out process can accelerate decision-making by taking an Agile approach. It is good for either many small decisions that need to happen quickly or one big decision that involves many stakeholders.

One of the most powerful lenses you can put on a problem is probability. In the next chapter, we will learn the fundamental ideas of probability including how to calculate and visualize them.

NINE

Superpower: Understanding Probabilities

If you are a fan of James Bond, you know this guy can calculate risk. He calculates it when he fights and he calculates it when he gambles, like in Casino Royal. Now imagine James Bond with no ability to assess probabilities and impacts. That would make for short and boring movies. In the real world, however, most professionals (e.g., accountants, engineers, and scientists) fall short of James Bond's near-superhuman ability to assess the odds and have a very hard time both accepting and assessing uncertainty. Hailing from a deterministic world of acceptable but outdated and useless formulae, these professionals, for reasons of efficiency, often reach for what they know will be accepted, even if wrong, rather than put in the extra effort to achieve better results using new methods. In fact, by reading this book, you have affirmed that you are willing to go the extra mile for a better answer, even if it takes some further effort to derive and explain.

For something to be random, it usually happens at an unexpected time and/or in an unexpected place. There is a known distribution of the frequency and severity of earthquakes, but THE when and where are only known to the gods. Of course, random events don't have to be as dramatic as earthquakes, but that takes nothing away from their impact. Every day, we are exposed to estimates and predictions of events that are likely or unlikely to happen, such as *"Will it rain?"* or *"Will there be a stalled car on the way to work?"* These questions translate into decisions such as taking an umbrella or taking the service road to avoid traffic.

While we can't predict most events with absolute certainty, we can try to estimate how likely they are to happen using probability. This is no easy task, given that humans have historically been bad at assessing and modeling probabilities in their minds. Several studies have proven that pigeons do a better job than people.[20,21] Some suggest this is due to humans' reliance on heuristics (pattern detection) and judgment. Pattern detection was certainly useful to those early humans who knew by experience that seeing the grass move in the field could often result in becoming burgers for a tiger. This innate capability to detect patterns has created a false illusion of control (as mentioned in *Dance with Chance: Making luck work for you* by Makrdakis, Hogarth, and Gaba, 2009). You often hear gamblers claiming they have a "system" based on their subjective experience with a pattern. **But in the end, the laws of probability always prevail over time.** Because human heuristics and judgments often don't do a very good job of assessing risk and probabilities, humans can reach out for tools such as range estimates, probability theory, statistics, and Monte Carlo simulation to help counter such biases.

---

[20] Roberts WA, MacDonald H, Lo KH. *Pigeons play the percentages: computation of probability in a bird. Anim Cogn.* 2018;21(4):575-581. doi:10.1007/s10071-018-1192-0.21
https://www.livescience.com/6150-pigeons-beat-humans-solving-monty-hall-problem.html,

[21] *Dance with Chance: Making Luck Work for You*, Oneworld Publications.

Every decision superhero should understand probability and uncertainty in a way that is both useful and practical for making decisions. In this chapter, we will cover some of the mathematical models and concepts related to uncertainty and probability that you will use, as the total Decision Superhero that you are, in your models to mimic real-life behaviors.

Uncertainty, risk, and other important distinctions

In the realm of decision science and system analysis, several key concepts play crucial roles in understanding and navigating the complexities involved. Though we use the words throughout the book, it is worth taking a moment to crystalize these fundamental probabilistic definitions. The following definitions are at the base of any decision modeling involving probabilities. In Chapter 10, we will explore how these concepts apply to decision-making under various risk, uncertainty, and conflict conditions.

Deterministic systems are systems or scenarios where we can predict the outcomes with complete certainty. Imagine a simple machine where you input a number, and it adds two to that number. The result is always predictable, always the same given the same input—there's a clear cause and effect without any room for deviation.

Unlike the predictable nature of deterministic systems, *uncertainty* is a lack of knowledge about the behavior or range of possibilities for a given variable and the system generating it. We can call this *ignorance* due to a lack of information. Claude Shannon, the father of information theory, famously said, *"Information is the resolution of uncertainty."* Without a weather forecast, you would be completely at the mercy of the weather. For example, wearing a raincoat when it's sunny will be hot and uncomfortable or not taking an umbrella on a rainy day will be wet and uncomfortable. While we might have predictions about tomorrow's weather, there's always a degree of uncertainty until the day it actually unfolds. So if you hear the weather channel predict a 30% chance of rain, they are saying that given the current weather conditions and measurements, compared to similar conditions, it rained three days out of 10. It is also well known that for years, they have padded their probability estimates by up to 15%![22]

Variability refers to the extent to which a set of data can spread out or cluster closely together. For instance, if we measure the heights of a group of people, we might find that most heights cluster around an average, but there are always outliers—some significantly taller or shorter than the average. This spread or clustering illustrates the concept of variability within the data.

Risk is the possibility of gaining something valuable or suffering harm or loss under uncertain or variable conditions. The variables are often known but not their end state. The classic example are games of chance. You know the rules and the probability of the outcomes but not the result until you play the game. You can win and you can lose. That is the dual nature of risk. Investing in the stock market is another more practical example of risk; there is risk because the value of your investments can *go up* or *down*, influenced by many factors whose ranges are not always known.

Project Risk focuses on predicting and suggesting mitigations relating to uncertain events or conditions that, if it occurs, could positively or negatively affect a project's objectives. An example might be the risk of a key supplier failing to deliver a critical component on time, which could delay the entire project and impact its success.

Given the importance of these concepts, we have condensed them into a summary table (Table 12).

---

[22] https://www.mcgill.ca/oss/article/environment/problematic-perceptions-probability-precipitation.

TABLE 12: SUMMARY OF DECISION AND RISK ANALYSIS.

Concept	Description	Example
Deterministic	A system whose evolution can be predicted exactly.	A calculator adding two numbers.
Uncertainty	Lack of knowledge about the behavior or range of possibilities of a variable.	Weather forecasting.
Variability	Describes how spread out or closely clustered a set of data is.	The heights of a group of people.
Risk	The possibility of gain or suffering harm or loss; danger under uncertain or variable conditions.	Investing in the stock market.
Project Risk	An uncertain event that could positively or negatively affect a project's objectives.	A supplier's failure to deliver a critical component.

The binary life of continuous probability

"It's an energy field created by all living things. It surrounds us and penetrates us; it binds the galaxy together." – Obi-Wan Kenobi

At the heart of decision and risk analysis is the concept of probability. We use probability to build decision models, run simulations, sample distributions, and express the results as a range of outcomes (also known as percentiles). To paraphrase Obi-Wan Kenobi, probability and uncertainty (in simulation and modeling) are like the force. It surrounds us, flows through us, and binds the universe together. Events chains are often connected by uncertainty, so given the central importance of this concept for everything that follows, let's explore this elementary concept and the axioms (rules in the universe) that support it.

Kolmogorov's probability axioms

"Probability expresses the likelihood of the occurrence of an event in a defined space." - A.N. Kolmogorov

*A probability is a **statement of belief** on how often something could happen given a certain number of repetitions.* When assessing the probability of an outcome, you are assessing how often it is true (1) versus false (0). These 1s and 0 are known as events. According to (Kolmogorov, 1956), a probability within a defined probability space should always follow four simple axioms:

1. The probability of an event is a non-negative real number.

2. The probability that at least one of the basic outcomes in the entire sample space will occur is 1.

3. Probabilities of outcomes in the same sample space must add up to 1. For example, if you have a 53% chance of heads, then you have a 47% chance of tails.

4. If A and B have no elements in common and are truly independent, then the probability of both events are additive. For example, the P(A) = 10% and P(B) = 30%, then the probability of either event occurring is P(A)+P(B) = 40%

Following these axioms is not very hard because they are a natural consequence of simple rules of logic. The ones to always keep in mind are three and four because the other two axioms are implicit. In the end, calculating probabilities is based on accounting for the binary (true/false) outcomes for each state under study.

How to estimate or define the probability of something

Probability is a statement of belief derived either objectively by counting how often something happens or subjectively based on opinions and fundamentals. Let us understand their types in the following sections.

Objective probability (using data)

Objective probability[23] is simply derived empirically by counting how often something happens, such as how many rainfalls (events) you can expect in a year (defined space). Another example is how many defects per 1000 parts made.

The key things to remember about objective probability are as follows:[24]

- There is always a numerator and a denominator, as probability is always a fraction.

- Expresses long-run frequency that an event will occur

- Has historical data to calculate and backup assumptions

- Can be assessed using known stochastic/chance processes

---

[23] Some may argue that objective probability is just data informed subjective opinion. Our definition mostly implies the availability of data and not it's completeness.
[24] A. N Kolmogorov, Foundations of the theory of probability, Chelsea Publications, 1956.

Probability Notation Cheat Sheet

- **P(A) = 10%** means that event A has a 10% or 1 in 10 chance of occurring.

- **P(A₁B₂)** is the notation for joint probability.

- **P(A₂|B₂)** is the notation for conditional probability. The notation now includes the "|" pipe symbol which means "given."

If we are lucky enough to have sample data, doing a little probabilistic accounting is possible. By this, I mean a little math. Let's explore how you can use simple data from the news to gain deeper insights. Imagine that one morning while reading the news on your tablet, you catch an article on your city's current employment numbers. The article was analyzing skilled workers and their employment prospects. Being a consummate decision superhero, your first reflex is to copy the table and run some simple numbers in Excel. Here are the July statistics being cited in the news.

TABLE 13: LABOR FORCE STATISTICS FOR THE BIG CITY.

Big City Labor Force Statistics, July 20xx

	Skilled Workers	Unskilled Workers	Total
Employed	366,000	246,000	612,000
Unemployed	26,000	62,000	88,000
Total	**392,000**	**308,000**	**700,000**

Surprisingly, using some simple math, this table can give us much more insight into joint, marginal, and conditional probabilities. These words sound scary, but relax—it is easier than you think.

Joint and marginal probabilities

The starting point is fairly simple and intuitive: calculate the percentages of each state against the total. This is very easy to do in a pivot table when *selecting percentages against the total option.*

TABLE 14: BREAKING OUT THE NUMBERS INTO PERCENTAGES OF THE TOTAL.

	Skilled Workers (B_1)	Unskilled Workers (B_2)	Maginal Probability of A
Employed (A_1)	52.3%	35.1%	87.4%
Unemployed (A_2)	3.7%	8.9%	12.6%
Marginal Probability of B	**56.0%**	**44.0%**	**100.0%**

So, for example, *employed* and *skilled* workers account for 52.3% of the total (366,000/700,000), or if we were to consider *unskilled* workers as a whole, including *employed* and *unemployed*, that works out to 308,000/700,000 or 44%.

Now consider that the total columns are named *marginal probability*[25] and not *total*. If we were stopping at calculating proportions, we would still call them *total*. Marginal probability is simply the proportion of a group (Employment or Skill Level in this case). So here, 87.4% of people are employed (Group A_1) and 12.6% are not (A_2). Therefore, the probability of being employed versus not is simply 87.4% versus 12.6%. The same applies to being skilled and unskilled.

Sometimes, we are more interested in the joint occurrence of two (or more) characteristics that define a specific sub-group.

Using fancy notation, we know that the joint probabilities of being *Unemployed* and *Unskilled* are written as $P(A_2B_2)$ and reflect a value of 8.9% of the total population.

Calculating conditional probabilities

Conditional probabilities usually center on a subset of the population. As with everything else with probability, we can easily calculate all our conditional probabilities using the concept of fractions. Conditional probabilities are also at the center of Bayes Theorem (Named after Thomas Bayes). This mathematical formula describes how to update the probability of an event based on new evidence or data. Because this approach accounts for changing information, it's a powerful tool for making informed decisions when faced with uncertainty.

---

[25] The use of the word marginal refers the fact that the calculations are in the table margins. Pivot Tables refer to these as total rows and columns.

Equation 1: Bayes Theorem

$$P(A|B) = \frac{P(B|A) \times P(A)}{P(B)}$$

Q. What is the probability that I am employed given I am skilled?
This is noted as $P(A_1|B_1)$. Notice that the notation now includes the
"|" pipe symbol which means "given." To calculate this, we take the
people who are both *skilled and employed divided by the total proportion of skilled workers* or simply:

52.3% ÷ 56.0% = 93.4%.

Using the joint and marginal probabilities for B (Table 14: Breaking
out the numbers into percentages of the total), we can easily parse
out the other probabilities by dividing the joint probabilities by the
row totals (marginal probabilities of B) in Excel.

TABLE 15: PROBABILITY OF EMPLOYMENT GIVEN A SKILL LEVEL (JOINT
PROBABILITY/ROW TOTAL).

Probabilites of A|B

	Skilled Workers (B₁)	Unskilled Workers (B₂)
Employed (A₁)	93.4%	79.9%
Unemployed (A₂)	6.6%	20.1%

If we approach the question from the other side and divide the joint
probabilities by the column total (marginal probabilities of A), we
answer a different but no less interesting question: *The probability of
being skilled given your employment status*

TABLE 16: PROBABILITY OF BEING SKILLED GIVEN YOUR EMPLOYMENT
STATUS.

Probabilites of B|A

	Skilled Workers (B_1)	Unskilled Workers (B_2)
Employed (A_1)	59.8%	40.2%
Unemployed (A_2)	29.5%	70.5%

What is the probability of being unemployed given that I am un-
skilled? Or $P(A_2|B_2)$ = 70.5%. Now, that wasn't so hard. It's amazing
how spreadsheets and tables make math easier.

Now, what about situations when data is not available? Subjective
probability methods to the rescue!

Subjective or opinion-based probability

In situations where few or no previous occurrences of an event are
known to have happened, we need to *express a probability as a proportion of occurrences based on a set of foundational beliefs or perhaps the modeler's personal experience.* Imagine someone asking
about the probability of Germany invading France in the next five
years. Both a challenging and seemingly obvious question. On the
one hand, since 1701, or during the last 323 years, the French and
Germans have gone to war (including both World Wars) nine times.
Objectively, we can calculate wars over years (9/323) and simply
derive an annual probability of 2.7863% of going to war for those
two countries. But do we truly think there is an almost 3% chance
of these countries going to war today? France/German collaboration has come a long way since 1945. Firstly, both countries are tied
at the hips economically through the EU, and attacking each other
would be Mutually Assured Destruction (MAD) economically for
them and the other member states. Secondly, the world's maturity
and reality pre-1945 and post-1945 are very different for historical

reasons we are all aware of. Let us agree that this changes things and that perhaps the historical probability will not reflect the future. This is where the modeler's opinion comes into play to inform the probability estimate.

It is a hard task to know beforehand whether the modeler's opinion is calibrated.[26] Therefore, his/her probability estimate can only be deemed reasonable because inferring otherwise would be very difficult. Unfortunately, the only caveat is that if the modeler is not calibrated or bases their opinion on the wrong set of foundational beliefs, then their probability assessment may not concur with reality—thus potentially leading to an erroneous decision. The latter situation is also very difficult to infer otherwise. To know if someone is calibrated, you can either run a barrage of tests and workshops, or you can, if the data is available, look at their estimate versus actual ratios.

The key things to remember about subjective probability are these:

- Measures the strength of personal belief

- Used when no historical data is available because the event is unique

- Prediction markets are useful estimation tools

- Probabilities should be updated as new information is available (Bayes Theorem[27])

- Uses naïve probability as a starting point

Naïve probability means 50%. This means that when no information is available to inform an opinion of an event occurring, the starting point is 50%, and research will serve to refine this number over time. The rationale behind this is the idea of the fair coin toss. If there are

---

[26] Calibration refers to the concept of training an estimator through various workshops and exercises on how to assess the correct amount of uncertainty.

[27] We will see later how Bayes Theorem offers a solution to incorporating subjective probabilities into our decision analysis.

more outcomes, such as 3, 4, or 20, then naïve probability is 1 divided by the number of potential outcomes.

How the 2016 US presidential election proved people don't get probability

On November 8th, 2016, the US lived through a transformational election. What is notable are the many errors on probabilities were made before and up to Trump's official election. Understanding that probabilities are, in fact, wagers or business bets is a good starting point. For example, Nate Silver proclaimed that Trump had a 28.6% chance of winning.

What did that mean? Well, if you ran the election 100 times, Trump would have won almost 29 times. You could also imagine 100 parallel universes and 29 of them elected Trump and 71 of them elected Clinton. That is a far cry from a lock. However, due to bias and media predictions, everybody was surprised even though Trump had almost a 1/3 chance of winning—and he did. The moral of the story is that just because the odds are in your favor, it does not mean you will prevail. Sometimes things go the other way.

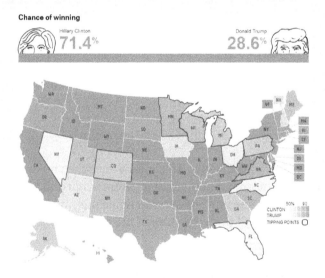

Figure 54: Results from www.fivethirtyeight.com.

The mathematics of expected value

Expected value is just a fancy term for a weighted average. That said, it does not detract from its central usefulness to the work at hand. In Excel, the SUMPRODUCT() function makes weighted averages a breeze.

If we want to calculate the expected value of a two-outcome situation, we multiply each probability by its respective value. It would look like this:

$$\textit{Expected Value} = p(\textit{winning}) \times \textit{return} + p(\textit{loosing}) \times \textit{loss}$$

We have set up a small Excel spreadsheet to experiment with. Assuming a 90% chance of winning $1 million and a 10% chance of loosing $100,000, that would result in an EV of $890,000.

	A	B	C	D	E	F	G
1		Analyzing win/loss scenarios using Probabilistic Expected Value					
2		Eric Torkia, Decision Superhero, 2024					
3							
4			Outcome	Prob.	EV		
5		Return	$1,000,000	90%	$900,000		
6		Loss	($100,000)	10%	($10,000)		
7							
8		EV			$890,000		
9		EV = (Prob. winning x Return) + ((1-Prob. winning) x Loss)					
10							
11		Address	Formula		Value		
12		D5	=1-D6		0.9		
13		E5	=+C5*D5		$900,000.00		
14		E6	=+C6*D6		($10,000.00)		
15		E8	=SUMPRODUCT(C5:C6,D5:D6)		$890,000.00		
16							
17							
18							

Figure 55: Simple expected value model.

We can get a payout table if we calculate the EV by varying the loss probability and the max loss. Such an analysis aims to get a sense of the win scenarios. *Risk-averse* decision-makers will be more preoccupied with all the scenarios, with a probability of losing that is < 50%. *Risk-neutrality* will consider the bet with a 50% value, and someone who is *risk-seeking* will still consider the bet in light of a loss probability > 50%.

SCENARIO ANALYSIS

Loss Scenario	Probability of loosing							
	99.50%	75%	50%	25%	10%	5%	1%	0.5%
$0	$5,000	$250,000	$500,000	$750,000	$900,000	$950,000	$990,000	$995,000
($10,000)	($4,950)	$242,500	$495,000	$747,500	$899,000	$949,500	$989,900	$994,950
($100,000)	($94,500)	$175,000	$450,000	$725,000	$890,000	$945,000	$989,000	$994,500
($250,000)	($243,750)	$62,500	$375,000	$687,500	$875,000	$937,500	$987,500	$993,750
($500,000)	($492,500)	($125,000)	$250,000	$625,000	$850,000	$925,000	$985,000	$992,500
($750,000)	($741,250)	($312,500)	$125,000	$562,500	$825,000	$912,500	$982,500	$991,250
($1,000,000)	($990,000)	($500,000)	$0	$500,000	$800,000	$900,000	$980,000	$990,000

Figure 56: Calculating expected value using loss probabilities and loss outcomes.

Graphing out the scenarios is equally interesting. It paints a simple but straightforward picture: as your probability of losing increases, your expected value gets closer and closer to a terminal value that is 100% of the loss.

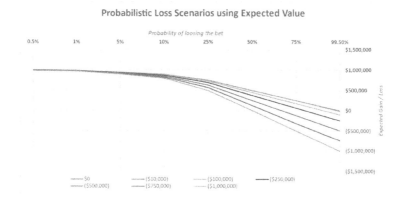

Figure 57: Comparing loss scenarios using Expected Value.

Analyzing expected value means, in its simplest form, that if *you take the bet repeatedly, perhaps into infinity, you would expect an average return of x.* The truth is that one of the predicted outcomes will occur and we often only get one chance to make the bet.[28] *Either you win or you lose. Either it happens or it doesn't.* These are extremely important ideas to keep in mind when thinking about EV.

---

[28] Monte-Carlo simulation is used to address these types of problems easily and effectively. We cover it extensively in book 2.

Fortunately, from a decision-making perspective, the expected value is a solid starting point for establishing a decision. The approach of weighting outcomes using probability adds a new layer of understanding and allows detailed analysis using decision trees.

Visualizing probabilities and decisions as trees

Decision trees are a graphical tool used in decision analysis to visually and explicitly represent decisions and decision-making. A decision tree consists of nodes and branches. Each node represents a decision point, chance event, or outcome, and the branches represent each node's possible choices or outcomes.

There are typically two types of nodes: decision nodes (usually depicted as squares) signify points to make a decision, and chance nodes (shown as circles) indicate points where a random event will occur with a certain probability. The tree branches coming out from these nodes lead to different outcomes or further decisions. Each outcome has an associated value or cost, allowing decision-makers to systematically evaluate and compare different strategies. Following the tree from the root to the leaves allows one to see all possible decision paths and outcomes, making it essential for complex decision processes.

In addition to decision problems, probability and decision trees are used in machine learning and configuring business rules in an enterprise system, credit scoring, and market segmentation, just to name a few modern examples and why it's important to be comfortable with these ideas.

Decision trees are our friend if we want to make a simple but important decision, such as to lease a new car or buy out the current lease. Notwithstanding the preference for novelty or new features, you need to consider what makes the most sense financially.

The second layer of decisions relates to warranty options. You can get

a new vehicle with the standard warranty or for $5000, you can pur-
chase an extended or premium warranty that gives you a lot more
coverage. In the same vein, if you were to buy out your current vehicle, you could ask the dealership to give you an extended warranty
that would also cost $5000.

*What is uncertain is how reliable the car will be during the next cycle
of ownership, which we can model as chance events.* For example, in
the car that you currently own, you have a good sense of whether or
not it's reliable, whereas, with a new car, you are confronted with no
information at all, meaning that it could be a lemon or it could be the
best car you've ever had.

While reading the decision tree, notice the square nodes reflect decisions, such as *leasing a new car, keeping the existing one,* or *buying an
extended warranty.* The round nodes define chance outcomes with
assigned probabilities such as *lemon, reliable,* or *failure.* All the math
revolves around expected value calculations that multiply the probability by outcome.

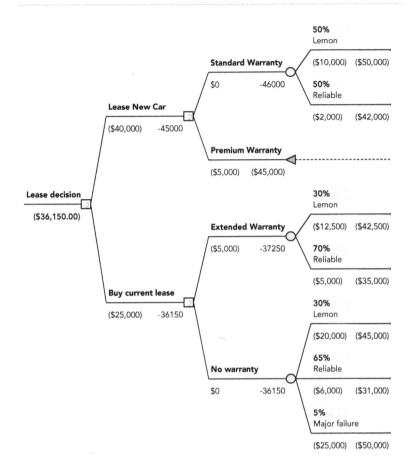

Figure 58: Decision Tree Example.[29]

What does this decision tree tell us? Firstly, if we assume the expected value of things (which is a big assumption and the only way the math works without simulation), the financial answer to your decision would be to *buy-out the current lease without a warranty.*

That is the best overall decision but if the car is a lemon, that decision would still cost $45,000 or the same as leasing a new car with a premium warranty!

---

[29] Decision tree developed in Excel using TreePlan (www.treeplan.com).

Now, let's consider that for an expected $1,100 difference in value, we get an extended warranty that would minimize the uncertainty of the costs over time. By selecting a premium warranty on a new car or an extended one on an existing one, you are willing to pay extra for peace of mind, knowing that most mechanical eventualities are covered.

Another side of you is willing to consider the possibility that nothing will happen and that spending the extra money is not worth it. You might even ask yourself the odds of a repair exceeding the warranty option price.

Ultimately, the right answer for you might depend on preferences that abstract from cost considerations such as novelty, peace of mind, familiarity, color, and finish.

We now know the power of probabilities and decision trees to better lay out our options, calculate expected outcomes, and even suggest a preferred course of action.

> *There are many ways to slice and dice an answer, but the best one truly depends on your needs.*

The good news is that there are rules you can apply to help select the best option based on risk attitude and the decision-making paradigm.

Are in-store extended warranties worth it?

The short answer is no when it comes to consumer electronics. In many cases, consumers find
that extended warranties for electronics are not
worth the extra cost, especially for less expensive
items. However, an extended warranty can be a
worthwhile investment for high-cost items with
expensive repair costs, or for those who value the
reassurance of having one.

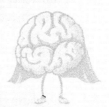

Elements to consider when evaluating extended warranties:

- Is the Cost of Warranty versus Cost of Product reasonable.
  Have a sense of the cost to repair the product, such as if it is a
  fraction of the product and less than the warranty

- Research the product reliability to see if they are prone to failure. Check the manufacturer's warranty to see if more coverage makes sense

- Does your Credit Card offer additional coverage? Some premium cards are known to automatically double the manufac-
  turers warranty. Like other insurance, they work using a claims
  process.

- How you use or intend to use the product may require additional coverage or void your coverage.

- Some warranty terms and conditions make using the coverage
  impractical, such as not having a guaranteed turn-around time,
  such as leaving your computer at the store for three weeks for
  the repair.

- Personal Peace of Mind

Key takeaways

- Probability is a statement of belief that is either informed using objective data or personal belief.

- Events happen or they don't. Either the bridge falls when hit or it does not.

- Calculating probabilities is easy with a spreadsheet and pivot tables.

- Expected value is a weighted average. It represents the long-run average by multiplying probability by impact.

- Probability trees make understanding things much easier

With a better understanding, we can now use these ideas and decision rules to improve our decision-making. These methods set the basis for analyzing more sophisticated analyses, such as Monte-Carlo simulations.

TEN

Probability and Decision Theory

S tatistical decision theory is a framework that combines statistical methods with decision-making under different conditions, providing a systematic way to make choices based on available data. It integrates concepts from statistics, economics, and operations research, focusing on optimizing decisions when outcomes are uncertain.

We can trace the roots of statistical decision back to the 18th century with the work of mathematicians like Thomas Bayes and Pierre-Simon Laplace, who laid the groundwork for Bayesian probability theory (*remember the formula from the previous section on conditional probability?*). Though Bayes is often credited for his contributions to the field of statistics, they would never have been had it not been for fellow mathematician and friend Richard Price, who posthumously edited and completed the famous article presenting Bayes ideas.

Statistical decision theory as we know it today matured in the mid-20th century, principally during the Second World War. In the

1940s-1950s, Abraham Wald (who we met in Chapter 1) laid the foundation of modern statistical decision theory with his work on sequential analysis and the theory of statistical decision functions. Wald introduced the concept of decision functions and risk, formalizing the process of making decisions under uncertainty.

Leonard "Jimmie" Savage (father to my friend and colleague, Sam Savage, who wrote the *Flaw of Averages*) expanded on Wald's work by formalizing the idea of subjective probability and utility, emphasizing personal beliefs and preferences in decision-making.

Savage's most notable contribution is his 1954 book, *The Foundations of Statistics*, which profoundly influenced the field of statistics and the philosophy of decision-making. In this work, Savage rigorously developed the concept of subjective probability and utility, laying out a framework for decision-making under uncertainty known as the *"Savage axioms."* These axioms formalize the idea that a utility function can represent individual preferences among uncertain prospects and that these preferences should adhere to certain consistency principles.

The Savage axioms

The Savage axioms are a series of conditions that a rational decision-maker's preferences should satisfy when choosing among different actions under uncertainty. These axioms serve as the foundation for the expected utility theory in a subjective setting, where probabilities are not known objectively but are instead personal judgments. The main axioms include:

TABLE 17: THE SAVAGE AXIOMS.

Axiom	Description	Example
Completeness	For any two actions A and B, either A is preferred to B, B is preferred to A, or the decision-maker is indifferent between them.	Choosing between tea (A) and coffee (B): you prefer tea, prefer coffee, or have no preference.
Transitivity	If A is preferred to B, and B is preferred to C, then A is preferred to C.	If you prefer tea over coffee, and coffee over milk, then you should prefer tea over milk.
Sure-Thing Principle	A decision-maker's preference between A and B should not change if the decision-maker would make the same choice in either event E or its absence.	Whether it rains (E) or not, if you would carry an umbrella, your decision should not depend on the weather forecast.
Non-Triviality	Not all actions are considered equal; there exists some variation in preferences.	You have a preference for some ice cream flavors over others, not all flavors are equally preferred.
State-Independence	Preferences between outcomes should not depend on the state of nature that leads to those outcomes.	Your preference for winning $100 should not depend on whether it comes from a lottery or a gift.

Axiom	Description	Example
Dominance	If one action is better than another under some circumstances and no worse under all other circumstances, it should be preferred.	If Job 1 offers a higher salary than Job 2 in both a booming economy and a recession, Job 1 should be preferred.
Continuity	If A is preferred to B, and B is preferred to C, then there should be a mix of A and C that is equally desirable as B.	If you prefer a beach vacation (A) over a city vacation (B), and a city vacation over a mountain retreat (C), there should be a mix of beach and mountain retreat that you find as appealing as the city vacation.

Along with the contributions of Bernoulli, Von Neumann & Morgenstern, Kolmogorov, and Wald, these axioms are foundational in decision theory, particularly in the context of making decisions under uncertainty. They aim to characterize rational behavior by providing conditions that preferences should satisfy to be considered consistent and rational. The examples given for each axiom illustrate practical applications or manifestations of these principles in everyday decision-making scenarios.

It is obvious that we require mathematics for analytics. *Applying math to physical things is often straightforward. But applying math to the strength of preference and to risk attitudes is less clear.* Let us look at statistical decision theory, beginning with the fundamental concept of how useful something is, which put simply, *is defined by its relative value to the person owning it.*

Thinking in terms of usefulness and preference

The key portions of the story of the development of analysis for decision-making are largely catalyzed by the development and adoption of risk analysis and the broader concept of decision-making. The development of *utility theory* is credited to *Daniel Bernoulli*, who states that **the value of an item is not derived from price, but instead how useful the item will be—the utility of the item.**

Put simply, a dollar is worth more to a beggar than a millionaire. For each additional dollar the beggar gains, the "usefulness" or necessity for that dollar diminishes, and more so for the one thereafter. If the process of gaining additional dollars continues until he is a millionaire, the incremental value of each additional dollar beyond a certain point is almost nothing. What's a dollar or even $1000 worth to a multi-millionaire? *Another more succinct way of summing up utility is the idea of risk versus reward or, as we say, "Bang for the buck!"*

According to author Peter Bernstein, Bernoulli was the first to measure something uncountable, and he credits him with heavily influencing psychology by providing a perspective on rational thought.

Bernoulli's proposition highlights how individuals' *risk attitudes impact decision-making, particularly in uncertain scenarios.* When faced with choices involving uncertain outcomes, people often opt for a certain outcome, even if it falls notably below the expected value associated with the uncertainty of winning the bet.

John von Neumann provided the next significant development in utility theory in the 1940s. He collaborated with Oskar Morgenstern to write *The Theory of Games and Economic Behavior*, which provided an axiomatic basis for utility theory and influenced the subsequent development of the field of decision analysis. He also developed game theory, where we determine outcomes by players' decisions rather than uncertainty.

Though we consider utility critical when framing the decision, for the purposes of this book, we shall assume that the preferences of the decision-maker (or decision scientist doing the analysis) among the various possible alternative outcomes can be analyzed in a straightforward manner. Because problems of utility theory are concerned with analyzing and understanding people's risk preferences, they can make for very difficult concepts when considering multiple individuals.

Therefore, to keep things manageable when analyzing decisions, we shall assume the following:

- We can measure or evaluate all the outcomes on a single scale, usually money, but it can be as varied as the context requires. For example, hospital beds, people saved, accidents avoided, and so on.

- The decision-maker wants to maximize their return or minimize their costs (monetary or otherwise).

- The decision is subject to some constraints.

- Not all outcomes are equally desirable.

- Decision-makers are aware of all the relevant courses of action (strategic responses). Because this is not always true, it becomes an assumption.

- The preference of alternatives is clear in the mind of the decision-maker. Once again, this is not always true and can be considered an assumption also.

Understanding one's risk attitude and valuing alternatives are crucial in making informed decisions amidst uncertainty. Individuals' risk attitudes significantly influence their choices, whether seeking greater gains, avoiding potential losses, or prioritizing expected values, as elucidated through mathematical analysis. So, let's take a few minutes to explore the mathematics necessary to calculate utility.

The mathematics of utility

Utility is yet another way to compare the relative attractiveness of options. This is expressed as a scalar value when evaluating a specific option and as a curve when thinking about the problem as a continuum.

When most people think of utility, they think of a utility curve. To plot one out, we need to calculate the utility at different points and fit a curve. The variable that controls the shape of the curve is called a *preference probability*. In simple terms, if you prefer odds that are better than chance when betting, you can be referred to as being risk averse. If you prefer odds where the potential reward outweighs the risk of losing, you would be known as risk seeking. If you are ok going with the expected value of the bet, you are risk neutral. *We will cover this in further detail in the next section and invite you to follow along with the example files.*

TABLE 18: PREFERENCE PROBABILITY.

Risk Attitude	Preference Probability
Risk Seeking	<0.5
Risk Neutral	0.5
Risk Averse	>0.5

The next piece of information are the upper and lower bounds of a problem. For example, an organization may consider projects that lose $100,000 to have a utility of 0 and those that make $800,000 a utility of 1. With these two important but simple pieces of information and three equations, we can plot the curve or calculate the utility of any alternative under study.

The **r** equation is to convert the probability into odds:

$$r = \frac{Pref.Probability}{1 - Pref.Probability}$$

The second formula calculates **alpha**, which shapes the curve. If the preference probability is set to 0.5, **alpha = 1**. In all other cases, this function should be used:

$$alpha = \frac{Pref.Probability^2}{2 \times Pref.Probability - 1}$$

We need to use the **r** and **alpha** terms we calculated earlier to calculate the utility function:

$$u(x) = alpha \times \left(1 - r^{\frac{-2(x-LowerBound)}{UpperBound-Lowerbound}}\right)$$

Once again, if we are in a risk-neutral situation using a preference probability of 0.5, then we need to revert to the linear form:

$$u(x)_{p(0.5)} = \frac{x - LowerBound}{UpperBound - LowerBound}$$

Expected Utility is simply the average of all the utilities in a problem.

Using the example values (-100,000 |10%; 800,000 | 90%) and a preference probability of 0.6, we plotted the following curve representing the utility curve for evaluating projects for an organization or perhaps the alternatives in a decision:

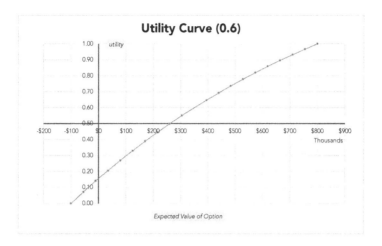

Figure 59: Utility curve with a preference probability of 0.6.

The best way to learn something, especially utility and risk preferences, is by telling a story and analyzing the lessons (the MBAs call this a case study).

Is a 1 in 10 chance a big probability?

Imagine you get a call from the host of the local *Morning Maniacs* radio show. They tell you that if you hurry and get to the station within 30 minutes, they will give you $10,000. Consider that it would take you at least 45 minutes to get there *safely* because of traffic, the time of day, and various small hiccups that could happen along the way. If you knew that you could make it there in 30 minutes or less but had a 10% chance of dying on the way, would you go? If not for $10,000, for what number?

$$Expected\ Utility = \frac{\Sigma(u(x_i) \times p_i)}{n\ outcomes}$$

What if the risk is not one of death but one of accident. The accident could incur a damaged car, broken neck, or both. It's hard to tell, and if you have to make a quick decision, all these things are flying around in your head, desperately trying to decide whether to take the trip and hopefully collect the money. What am I willing to wager to make $10,000? My life, my car, my life prospects?

One could cynically summarize this dilemma as *"Everyone has got a price."* This, my fellow decision superhero, *is the basic underpinning idea behind utility.*

This cynical view suggests that for every risk, a monetary amount would compel an individual to accept it, reflecting their personal utility curve. However, when it comes to life-threatening risks, in most cases, this saying reaches its limit as many people would find no sum of money worth the risk of death, highlighting the personal nature and variability of human utility. But then again, there are still some out there who would do risky things anyways. History is full of stories of risk seekers. Virtuous examples mostly revolve around people who put others well being above their own (such as first responders) while others are seeking thrills or perhaps just dumb. You know the old saying about there being a thin line between stupidity and bravery? Some notable examples include:

- *Franz Reichelt's Parachute Suit Failure (1912):* Poor Franz, an innovative tailor who designed a parachute as a novel piece of safety equipment. So convinced in his invention, to test it, he jumped off the Eifel Tower, and well, let's just say his design was flat out wrong.

- *Evel Knievel's Snake River Canyon Jump (1974):* Evel Knievel, an American stunt performer, attempted to jump the Snake River Canyon, which is a ¼ mile long in Idaho on a steam-powered rocket called the Skycycle X-2. Unfortunately, the stunt failed because the parachute deployed prematurely, causing the rocket to drift down into the canyon rather than across it. Knievel survived with only minor injuries, but in the end, the Skycycle was really a ground cycle in disguise.

- *Karel Soucek (1984-1985):* Soucek is one of the last to survive a plunge over Niagara Falls in a custom-built barrel.

Alas, seeking to replicate his success a year later, Soucek died attempting to repeat the stunt at the Houston Astrodome by being dropped from a height of 180ft into a water tank. The stunt tragically failed because the barrel hit the side of the water tank meant to cushion his fall.

- *Robert Overcracker (1995):* As a publicity stunt to promote the cause of homelessness, Overcracker decided to ride his jet ski over the brink of Horseshoe Falls with a parachute. Tragically, his parachute did not open and posthumously, Overcracker ended up promoting better parachutes instead.

Cynicism aside, depending on the answer that came to mind when wondering if you would make the trip to the radio station, puts you into one of three basic risk attitudes. Since everyone is unique and comes to the table with different experiences and knowledge, they may have very different risk attitudes to your own. Recognizing this allows the decision superhero to avoid making false assumptions about how people see the world.

Risk attitudes fall into three categories: *"I like risk," "risk scares me,"* and *"I don't care either way."* More formally, they are known as risk seeking, risk averting, and risk neutral, respectively. Let's use the Morning Radio example to illustrate, but instead of death, let's assume we would either total the car, get a ticket, or both. In this case, our loss would be $4,000 for the car (it's a beater) and $1,000 for the ticket for reckless driving, resulting in a max total loss of $5,000. Doing the math, the EV is $8,500.

	Value	Probability	EV
Return	$10,000	90%	$9,000
Loss Potential	($5,000)	10%	-$500
EV			$8,500

Figure 60: Calculating the EV[30] for the Morning Maniac Problem.

---

[30] EV = (Prob. of Win x Return) + ((1-Prob. of Win) x Value of Life)

The relative usefulness of this wager to the better depends on, as Bernoulli put it, their risk attitude.

Risk seeking

Risk seeking is characterized by a preference for outcomes with higher variability, even at the expense of lower expected values. Mathematically, a risk-seeking individual is willing to accept a gamble if its potential payoff outweighs the certain outcome. This includes gamblers, stock traders, and stunt men. **These individuals would set their *preference probability < 0.5*.**

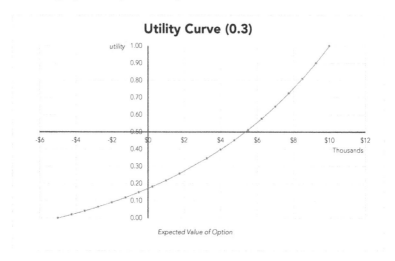

Figure 61: Example of Risk Seeking Utility Curve.

A risk-seeking person might be more inclined to take the challenge, valuing the thrill of the gamble or the monetary reward more highly than the potential risk. However, the extreme risk of death might still be a deterrent unless the reward is significantly higher. *The utility for an EV of 8,500 with a preference probability of 0.3 is 0.81.* Risk seekers would be interested in this bet but less so than risk averse or risk neutral people.

Risk averting

Risk aversion reflects a preference for outcomes with lower variability, even if it means accepting lower expected values. Mathematically, *a risk-averse individual prioritizes minimizing potential losses over maximizing potential gains.* **These individuals would set their preference probability > 0.5.**

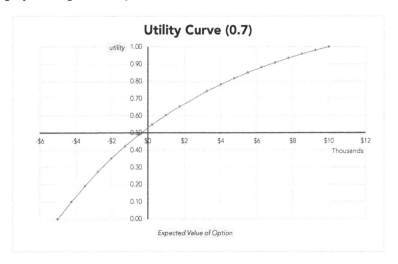

Figure 62: Example of Risk Averse Utility Curve.

Most people naturally dislike taking risks, especially when their health or life is on the line. A risk-averse person might not attempt this dangerous dash for $10,000, valuing their safety over the monetary gain. For them, the utility (satisfaction or happiness) gained from staying safe outweighs the utility of the money. *The utility for an EV of 8,500 using a utility curve with a preference probability of 0.7 is = 0.96.* Risk averting people would see this as a good deal.

Risk neutral

Risk neutrality is the indifference towards variability in outcomes, focusing solely on maximizing expected value. Mathematically, risk-neutral individuals evaluate choices solely based on their expected values, without consideration for risk. **These individuals would set their** *preference probability = 0.5.*

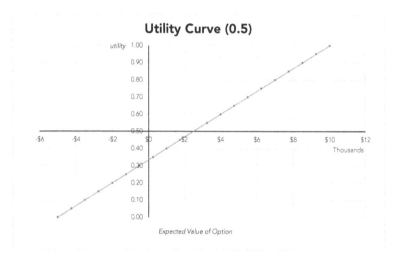

Figure 63: Example of Risk Neutral Utility Curve.

A risk-neutral person evaluates options based on expected outcomes without weighing the risks heavily. They might calculate the expected value of the gamble (in this case, factoring in the 10% chance of death) and decide purely on that basis, possibly still avoiding the risk for $10,000 due to the extremely high stakes. *The utility for an EV of 8,500 using a utility curve with a preference probability of 0.5 is = 0.9.*

TABLE 19: COMPARING UTILITY SCORES BY RISK ATTITUDE.

	Risk Averse	Risk Neutral	Risk Seeking
Preference Probability	0.7	0.5	0.3
Utility	0.96	0.9	0.81

In this summary table comparing risk attitudes, we can see that the idea of running across town to get the $10,000 prize is a good deal if you are willing to lose $5,000. But what happens if $5,000 is not there to lose? You would be very risk averse, perhaps to the point of forgoing the bet altogether. Ironically, distress can be a motivator to take on more risk and not less. If someone needs the money for an emergency like a medical bill or a mortgage payment, they may consider risk the only solution. When you get anxious, crazy ideas start sounding good. I know more than one bank robbery that started this way.

The principle of certain monetary equivalence

Certain monetary equivalence (CME) is the guaranteed return you would accept today instead of risking a potentially higher but uncertain return. It represents the specific amount of money you view as equally desirable to the uncertain prospect of winning more.

For example, imagine you're offered a choice between receiving $100 guaranteed right now or a 50% chance to win $200 tomorrow. How much would it take for you to walk away without making the bet? If you are satisfied with $150 guaranteed today as you would be taking the gamble for the full $200, then $150 is your certainty equivalent for this particular deal.

Exploring the concept of the Certain Monetary Equivalent (CME) unveils various approaches.

Financial approach to CME (f-CME)

The financial approach focuses on applying discount rates to evaluate future cash flows. This method relies on the principle of the time value of money, using a discount rate to adjust future cash flows to their present value. The chosen discount rate reflects the risk associated with the cash flows, the opportunity cost of capital, and expected inflation.

$$CME = \frac{Expected\ Value}{(1 + risk\ free + risk\ premium)}$$

The output of this process is the certain monetary equivalent (CME), which represents the lump sum present value of these future cash flows. Using Excel, we put together a simple example winning $1 million 70% of the time versus a 30% chance of losing $500,000 (Figure 64).

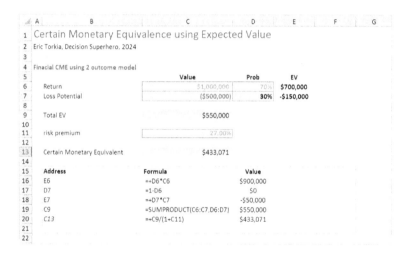

Figure 64: Financial CME using a 2-outcome model.

Using a 27% risk premium, the CME is $433,071, meaning that in this perspective, one should be willing to accept a lower amount today to offset the risk of tomorrow.

This technique is predominantly used in areas such as capital budgeting and investment analysis, where the goal is to maximize financial value.

Economic approach to CME (e-CME)

Contrastingly, the economic approach employs utility functions to determine a CME. This method assesses outcomes based on the utility or satisfaction they provide to an individual, rather than their mere financial value. Utility functions translate monetary outcomes into a measure of utility, reflecting the subjective well-being or satisfaction derived from different levels of wealth. Using the same Excel model as above, we incorporated some extra math to calculate the utility of an outcome, in this case EV, against its implied curve.

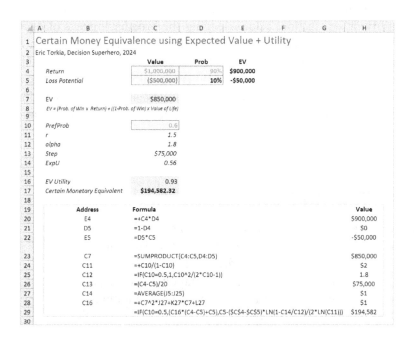

Figure 65: CME using expected value and utility.

Comparing apples to apples, the Financial CME is $433K, while the economic perspective is significantly more conservative at $194K, which is less than half! *The economic perspective is useful in setting the absolute lower bound for a decision-maker based on his risk preference, while the financial perspective seeks to maximize the return in the face of risk.*

The key distinction between these methodologies lies in their approach to valuing uncertain outcomes. The financial method adjusts for risk externally through the discount rate, effectively demanding a risk premium for future uncertainty. In contrast, the economic approach integrates risk preferences directly into the decision-making process via the utility function, providing a more nuanced view of how risk influences choices without necessitating external adjustments.

While the financial approach aims to maximize wealth from a market-based perspective, the economic method seeks to maximize utility or satisfaction, acknowledging the broader impacts of decisions on individual well-being. As such, the financial approach is most suited to business and investment decisions where quantifiable cash flows and market considerations prevail. Meanwhile, the economic approach offers broader applicability, suitable for analyzing decisions under uncertainty involving subjective preferences and non-financial outcomes.

GreenGear Super-Widgets

Imagine a scenario where Bob, the pioneering founder of GreenGear Corp, seeks your counsel on launching a groundbreaking product designed for super-widgets. This product promises to catapult GreenGear into a previously unreachable market, dismantling long-standing entry barriers.

The following Sunday, while golfing, Bob discusses this venture with his friend (and sometimes competitor) Jim, the innovative force behind Blue Dream Manufacturing Inc., a notable player in the widget industry. Bob excitedly shared the news about the super-widget and how they planned to compete in the same market.

Jim explained that widgets are a fierce and competitive market, and you really need to know what you are doing. Amid their conversation, Jim proposes an alternative strategy: rather than entering the competitive fray, GreenGear could license the super-widget patent to Blue Dream Manufacturing for a term of five years, equivalent to the first year's revenue. Jim underscores the widget market's cutthroat nature, emphasizing the necessity of industry acumen. Bob listened, suddenly concerned that maybe they made a big bet and might lose!

Sensing Bob's concern, Jim says, "Look Bob, we have known each other a long time and you know I am being straight and trying to help. Tell me what you think is a reasonable estimate for your first year and if it makes sense to us—let's do it!"

Monday morning, Bob convenes with you, Ralph, the marketing director, and Dave, the finance chief, to outline four possible revenue scenarios: high, medium, low, and loss. Employing a straightforward decision tree, you incorporate these scenarios with associated values and probabilities to compute the expected value of the project's revenue potential.

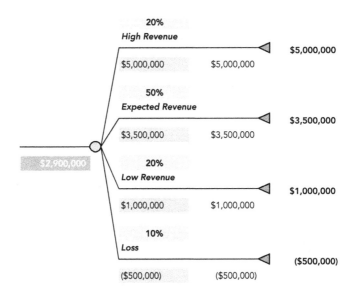

Figure 66: Calculating the expected value using multiple outcomes.

Bob turns around and asks you to do the math so the team can consider it. Before getting started, you explain that the decision is a personal one that resolves around risk preferences. When asked, Dave was risk averse and considers a bird in hand to be worth two in the air while Ralph, being the consummate marketing optimist, thinks we should hold out for more. On the other hand, Bob is very matter of fact when it comes to transactions and considers himself the Switzerland of risk—totally neutral!

CME using multiple weighted outcomes

	Value	Prob	EV	Utility
High Revenue	5000000	20%	$1,000,000	1.00
Expected Revenue	3500000	50%	$1,750,000	0.80
Low Revenue	1000000	20%	$200,000	0.36
Loss	-500000	10%	-$50,000	0.00

Preference Probability	0.60	r	1.50
Risk-free rate	3.38%	alpha	1.8
risk premium	10.00%	ExpU	0.539780919

Total EV	$2,900,000
f-CME	$2,557,770
e-CME	$1,917,910

	Neutral	Risk Seeking				Risk Averse
Risk Premium (incl. risk-free)	3.4%	5.0%	12.5%	25.0%	33.3%	50.0%
Risk Adjusted EV	$2,805,185	$2,761,905	$2,577,778	$2,320,000	$2,175,544	$1,933,333
Risk Price Adjustment	($94,815)	($138,095)	($322,222)	($580,000)	($724,456)	($966,667)
Premium over CME	$887,275	$843,995	$659,868	$402,090	$257,634	$15,424
CME	$1,917,910	$1,917,910	$1,917,910	$1,917,910	$1,917,910	$1,917,910

Figure 67: Calculating multiple risk scenarios.

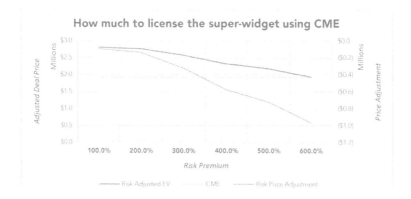

Figure 68: Adjusting expected value using a risk premium.

By plotting out the scenarios for the CME at different risk premiums, we can evaluate how much we are willing to accept based on a personal risk attitude. You suggest putting the values for each opinion into a table, knowing that ultimately Bob has the final say.

TABLE 20: COMPARING RISK ATTITUDES ON PRICE.

Scenario	Attitude	Premium	Adjusted EV
EV Sell Price	Unadjusted	0%	$2,900,000
Economic CME	Unadjusted	33.9%	$1,917,910
Dave	Averse	30.0%	$2,230,769
Ralph	Seeking	5.0%	$2,761,905
Bob	Neutral	3.4%	$2,805,185
Team Average		12.8%	$2,599,286
Premium over CME / negotiating room			$681,377
Risk adj. to EV sell price			**($300,714)**

Bob values his team's opinion and suggests that all opinions carry the same weight. Being the principal decision-maker, Bob can:

- Pick the scenario of his choosing regardless of the opinion or data others may have.

- He could decide to ask for more than *the Expected Value* (a scenario still outside of the analysis) because he feels that the product would be a huge success.

- Bob can also take a consensus position and simply average out the estimates to roughly $2.6M (a $300,000 risk premium).

- Bob can propose the economic CME in a fire-sale move to get cash quickly at the smallest acceptable sell price.

In whatever case, it will be an amount that, if offered today, GreenGear Corp would be willing to forgo the next five years of revenue (and expenses) selling and distributing the super-widget independently.

But wait, there was a 20% chance of the product selling $5 million+ but that is also subject to many factors that are not all under the control of GreenGear. After calculating the expected value and senior management agreeing to a risk premium, Bob called Jim to propose a $2.6M licensing deal knowing he had the ability to negotiate another.

When Jim heard the offer, he was delighted because he suspected that with his network of distributors and manufacturing knowhow, he will make a killing. Bob is happy because he gets a certain $2.6M (or 35% over his minimum sell price) when he is not sure how successful the product will be in the future, even if he really hopes so. In the end, both parties are happy with the number and that is what counts.

We can achieve remarkable outcomes in decision analysis by employing straightforward mathematical approaches such as expected value, probability weighting, and certain monetary equivalence.

Decision-making paradigms

The essence of decision-making under any circumstances can be summarized as a selection between two or more courses of action/strategies. In statistical decision theory, there are five basic operating paradigms. Some are very straightforward, while others require more work to get a reasonable answer.

Decision-making under certainty

Often, we deal with situations of preference or constraint that are well known to us. Essentially, *all available options are known and pre-*

dictable, so making a decision relies entirely on the decision-maker's objectives. For example, coffee or tea with your eggs? Or, where are we going for lunch today—our favorite pizza or Chinese place?

Decision-making under complete ignorance

In this scenario, decision-makers *have no information on which to base their decisions*. Outcomes are completely unknown and unknowable. One course of action is to leap and hope for the best and another is to do nothing.

"Some would even question if this [decision making under complete ignorance] is a real decision-making situation in the first place and suggest that solutions to these types of decisions are more psychological or philosophical than statistical" (Richardson S. B., 1965).

Decision-making under risk or partial uncertainty

This type of decision-making involves known risks and underlying models and processes that are understood, thus allowing the calculation of the likelihood of various outcomes using statistical models. Games of chance, such as coin tosses, cards, or die, all fall under the rubric of deciding under risk. For example, we know that a fair coin has a ½ chance of being heads or tails, or 4/52 (1/13) to draw a king from a deck of cards. From a business point of view, the probabilities are known to some or all of the courses of action under study.

Investors often rely on historical performance data to assess the risks and expected returns of different investment options in finance. One more example is managing quality. Statistical quality control uses sample data and known processes to decide whether to accept a lot or return it to the supplier based on a sample of defective parts.

Decision-making under conflict

The states of nature are subject to the control of an adverse intellect. Decision-making takes into account adverse reactions and seeks to optimize decisions in response. The basis for this approach is squarely rooted in game theory, and although math shares similar approaches to decision theory, decision-making under conflict is beyond the scope of this book. However, the methods for decision-making under uncertainty will prove useful and may give some answers when confronted with this type of decision-making (Savage L. J., 1971).

Decision-making under complete uncertainty

Decision-making under complete uncertainty occurs when a decision-maker has to *choose among various alternatives without knowing the probabilities of the various outcomes.* This situation represents a high level of ambiguity because there's no reliable information available on the likelihood of the outcomes, making it challenging to predict the results of any decision. In such scenarios, decision-makers must rely on different decision-making rules or criteria to guide their choices. Here are several rules commonly used for making decisions under complete uncertainty:

Maximin (Minimax) rule

Maximin is for the risk-averse decision-maker. You evaluate each alternative's worst possible outcome and then choose the "least bad" worst outcome. Essentially, you're maximizing the minimum gain (or minimizing the maximum loss), hence "maximin" for gain-focused scenarios and "minimax" for loss-focused scenarios.

Maximax rule

Maximax appeals to risk-takers or optimists. Here, you consider the best possible outcome for each alternative and choose the one with the highest potential gain. It's the opposite of the Maximin rule, focusing on the most positive scenario.

Minimax regret rule

The *Minimax Regret rule* involves calculating the regret for not choosing the best course of action in hindsight and then choosing the alternative that minimizes the maximum regret. Regret is the difference between the payoff of the optimal decision and the payoff of the decision made. This approach aims to minimize potential feelings of remorse from a decision.

Equal Probability (Laplace) rule

Under the assumption that all outcomes are equally likely, the Laplace rule suggests calculating the average payoff for each alternative and choosing the one with the highest average. *This rule treats all unknown probabilities as equal, effectively distributing one's ignorance evenly across the outcomes.*

Hurwicz Criterion (Optimism Criteria)

The Hurwicz Criterion introduces a parameter α (alpha), which measures the decision-maker's optimism. By weighting the best and worst outcomes for each alternative according to α and $(1-\alpha)$, respectively, this rule allows for a personalized balance between optimism and pessimism in decision-making.

Savage's Criterion (Minimax Regret)

Similar to the Minimax Regret rule but distinct in its formalization by Leonard Savage, this approach focuses on minimizing the maximum regret one would experience from not having chosen the best alternative in hindsight.

These rules provide structured approaches to making decisions in the absence of clear probabilities, suggesting a path to navigate uncertainty based on their risk preferences, optimism, and strategic considerations.

Key takeaways

- The Savage Axioms propose a way to compare options.

- Utility can be summarized as a dollar is more useful to a beggar than to a millionaire.

- Your risk attitude will dictate which options are more attractive.

- The Certain monetary equivalent tells you the minimum you would be satisfied with given a probability and a set of outcomes.

- Rules exist to analyze decisions under complete uncertainty.

In the next two chapters, we will explore how to usefully apply these rules in conjunction with probability to gain new insights and potential avenues for the best decision.

ELEVEN

SuperPower: Calculating The Value of Perfect Information

One of the tools in decision theory is to assess the impact of having better information. Unfortunately, better information costs money, and there are limits in terms of time and economic feasibility for any analysis. In these cases, *the Value of Perfect Information (VOPI)* can help the decision superhero plan and budget for better data while considering what it is worth and how much is needed to make a difference.

 VOPI is structured using payoff matrices with values and probabilities. We will revisit our food truck and see how to use weather information to improve daily sales. Then, we will get fancy and further improve the accuracy of our analysis by applying Bayes theorem. *To make following along easier, all the analysis is available in the chapters example files.*

Calculating the impact of better information using simple probabilities

Recall the food truck problem from Chapter 8. We run a food truck in New York City and want to optimize our profits by figuring out what we need on any given day to make our customers happy. On a rainy day, history has proven that we sell more coffee and donuts, and on a sunny day, history shows that we sell a lot of ice cream and tacos. On days where we get it wrong, we can incur up to $1,000 in food waste because it's a fresh batch every day, and whatever is not sold is lost. This situation is also known as the *Newsboy problem*, coined after the fact that if a news boy did not sell all his newspapers that day, he could not re-sell yesterday's news and would take a loss for anything unsold.

Defining the naïve state (no information scenario)

If we frame the decision problem using a payoff matrix, we have two courses of action (also known as strategies[31]), two states of nature, and four outcomes. Putting the information in this form will allow simple matrix math in Excel to complete the job. In the old days, they had to do everything by hand!

---

[31] Courses of action, decisions and strategies are used interchangeably.

TABLE 21: FRAMING THE FOOD TRUCK DECISION PROBLEM OUTCOMES.

Outcomes *Course of Action*	*States of Nature*	
	Sun	*Rain*
Ice Cream + Tacos	**Win/Profit**	*(Loss)*
Coffee + Donuts	*(Loss)*	**Win / Profit**

$ Value *Course of Action*	*States of Nature*	
	Sun	*Rain*
Ice Cream + Tacos	$2,250	($1,000)
Coffee + Donuts	($1,000)	$2,750

In recent years, Excel has been upgraded to naturally accept cell ranges. This means you can easily multiply two matrices together in one formula without using a yesteryear's special array formulas. You can find an example of this in the book's support files.

TABLE 22: CALCULATING EXPECTED VALUE WITH NO INFORMATION.

Probabilities *Course of Action*	*States of Nature*		Total
	Sun	*Rain*	
Ice Cream + Tacos	50%	50%	100%
Coffee + Donuts	50%	50%	100%

Expected Value *Course of Action*	*States of Nature*		EV
	Sun	*Rain*	
Ice Cream + Tacos	$1,125	($500)	$625.00
Coffee + Donuts	($500)	$1,375	$875.00

Using the approach above, the *EV for Ice Cream and Tacos is (0.5) ($2,250)+0.5(-$1,000) = $625* and similarly, *Coffee and Donuts is (0.5)($2,750)+0.5(-$1,000)= $875.*

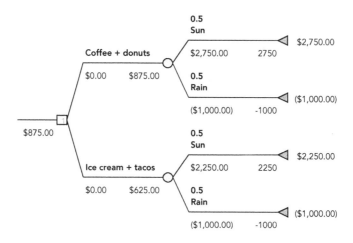

Figure 69: Presenting the Expected Value table as a decision tree.

If you total the EV for both courses of action, **a strategy with no information**, using naïve probability, **estimates a daily sales average of $1,500**. The analysis also suggests that *continually loading Coffee and Donuts will make you more successful in the long run.*

Updating analysis with probability estimates for weather

Using seasonal weather data, we can update our probability estimate and calculate an EV closer to what happens over a season. Using publicly available data, we roughly know that over the course of the summer in New York City, 70% of the time[32] it is sunny while 30% of the time it's raining. We get the following EVs for our two courses of action by updating our model.

---

[32] Time refers to days in the season. We looked at the proportion of days where it rained.

TABLE 23: EXPECTED VALUE WITH NEW INFORMATION ON WEATHER.

| Probabilities | States of Nature | | |
Course of Action	Sun	Rain	Total
Ice Cream + Tacos	70%	30%	100%
Coffee + Donuts	70%	30%	100%

| Expected Value | States of Nature | | |
Decisions	Sun	Rain	EV
Ice Cream + Tacos	$1,575	($300)	$1,275.00
Coffee + Donuts	($700)	$825	$125.00

At first glance, the new probability information indicates that the total EV is now $1,400, which is $100 less than before. *However, if we look at the individual EVs, the recommended (optimal) course of action has changed from Coffee + Donuts to Ice Cream + Tacos.* Assuming we switch strategies in light of the new information, we can calculate the value of improved weather data to revenue for an average day. To do this, we take the new optimal strategy with information and subtract the EV for the same strategy with no information.

Daily Improvement = Optimal Strategy with info **− Same Strategy** no info

In this case, we take *the new EV for Ice Cream + Tacos ($1,275), less the old EV of $625* and conclude that the new probability information could improve your daily sales for selling ice cream and tacos by up to $650!

Value of Switching = New Optimal Strategy with info **− Old Optimal Strategy** no info

Originally, the optimal strategy was to load up on coffee. However, considering the new probability information, *switching optimal* *strategies costs $1,275 (Ice Cream + Tacos) less $875 (Coffee + Donuts) = $400.* In both cases, the math suggests switching strategies is the way to go.

Calculating the value of perfect information

By accounting for weather probabilities, the simple optimal strategy is to just pack the truck with ice cream and tacos and hope for the best, knowing things will work out better in the end, generating $1,275/day on average. But this analysis also suggests that if you had perfect information, so a little fairy would tell you 100% of the time, with no error, what the next day's weather would be like, you could be making up to $2,400/day ($1,575+$825)! This is known as the Value of Perfect Information (VOIP)

$$\text{VOIP} = \text{Optimal EV}_{sun} + \text{Optimal EV}_{Rain}$$

Using the VOIP, we can also figure out an upper bound for the daily sales improvement we can hope to expect with better information, such as weather probabilities. Improved Prediction Potential (IPP) is calculated by taking the VOIP, in this case $2,400, and subtracting the EV for our default strategy (the best overall course of action), which is $1,275.

$$\text{IPP=VOI} - \text{Default Strategy EV}$$

The *improved prediction potential is $1,125 ($2,400-$1,275),* meaning that we can improve daily sales by up to that amount, given we have better information. The truth is that as decision superheroes, we know full well that the reality will be somewhere in between.

TABLE 24: SUMMARY OF VOIP ANALYSIS.

Decision Strategy for improving Seasonal Profit	Value
Optimal Strategy	Ice Cream + Tacos
Optimal Strategy Value - Average Daily Sales	$1,275.00
Value of Additional Probability Information per day	$650.00
Value of switching optimal strategies	$1,179.35
Value of Perfect Information	$2,400.00
Potential Improvement due to better forecasting	**$1,125.00**

Now, the management question, what would you pay on a daily basis to get a better prediction? If we rely on public sources such as the weather channel, we are still getting a probability for the day, and there is no certainty there, which is what is required for perfect information. In these cases, we can use Bayes' theorem to adjust our probabilities and account for forecast accuracy, wet bias or errors.

Using Bayes theorem for decision-making under risk

So far, we have seen the impact of accounting for simple probabilities on estimating the value of perfect information. Now, if we take this analysis a step further, we can also account for cloudy days using a conditional probability on the accuracy of forecasts.

The first step is defining what some might refer to as Type I and Type

II errors[33], and they would be correct. Applied in the context of a decision, we define a matrix of outcomes versus predictions.

TABLE 25: OUTCOMES VERSUS PREDICTION MATRIX.

New Information	Sun (b1)	Rain (b2)
Sunny Forecast (x1)	Sun/Sun	Rains when sun is predicted
Rain Forecast (x2)	Sunny when rain is predicted	Rain/Rain

We can assign probabilities to these outcomes:

TABLE 26: PROBABILITY THAT THE WEATHER FORECAST IS ACCURATE.

Probability of accurate forecast	Sun (b1)	Rain (b2)
Sunny Forecast (x1)	90%	20%
Rain Forecast (x2)	10%	80%

This table is very simple to interpret. 90% percent of the time when sun is predicted, it's sunny outside and 10% of the time we forget to bring an umbrella. Inversely, most weather predictions have what is called a wet bias, and for this reason, there is a 20% chance that we will get sun when rain is predicted. We will now use these conditional probabilities to adjust our seasonal estimate by calculating a posterior set of probabilities. Fancy talk to say that we will do a bit of matrix math. Hang tight, we'll get through this.

---

[33] A Type 1 error occurs when a true null hypothesis is incorrectly rejected (a false positive), while a Type 2 error occurs when a false null hypothesis is not rejected (a false negative).

Calculating the conditionals

The math, when broken out, is less daunting. You may want to open the example Excel file to follow the math. Trust me when I say it is easier that way.

Using Bayes theorem that we covered in chapter 9, let's take the probability of it being sunny when it is forecast to be **sunny or p(b1|x1)**

$$p(b1|x1) = \frac{Weighted\ Probability}{Weighted\ State\ of\ Nature}$$

Calculating the weighted probability is easy.

> = p(SunnyDay) × p(SunnyForecast)
>
> = 0.7×0.9
>
> =0.63

Calculating the weighted state of nature, concerns the total space.

> =p(SunnyDay) × p(SunnyForecast)+ p(Rain) × p(RainOnSunny)
>
> =0.7×0.9 + 0.3×0.2
>
> =0.6

Therefore, if we assemble the equation, the math is easy now.

> = $p(b_1|x_1)$
>
> = 0.63 ÷ 0.69
>
> =0.91

By repeating the calculation for **p(b2|x2) or Rain/Rain** scenario, we come up with 0.77 as our posterior probability. *So if the day's forecast is for sun, 91% of the time, it will be accurate. If the forecast is for rain, it will be accurate 77% of the time.* We have organized the results into a matrix.

TABLE 27: CALCULATING HOW OFTEN A PREDICTION IS ACCURATE.

Calculate conditionals	Sun/Sun	Rain/Rain
	P(b1|x1)	P(b2|x2)
Weighted Probability	63%	24%
Weighted States of Nature	69%	31%
Prob. Of what is predicted happens	**91%**	**77%**

P(b1|x1): Probability of a sunny day with a sunny day forecast prediction
P(b2|x2): Probability of a rainy day with a rainyday forecast prediction

We can define a complete matrix of posterior probabilities by taking the reciprocals. A reciprocal is when you take a 1 and subtract the probability to give you *the probability of the contrary outcome.*

TABLE 28: ORGANIZING POSTERIOR PROBABILITIES INTO A MATRIX.

Posterior Probabilities		
New Information	**Sun (b1)**	**Rain (b2)**
Sunny Forecast (x1)	**91%**	9%
Rain Forecast (x2)	23%	**77%**

Analyzing payoff scenarios

If we want to study the EV scenarios on a sunny day, we multiply the payoff matrix in Table 21 with the probabilities for a sunny forecast (91%/9%) in Table 28.

TABLE 29: EXPECTED REVENUE ON A SUNNY DAY.

EV on a Sunny Day			
Course of Action	**Sun (b1)**	**Rain (b2)**	**Total**
Ice Cream + Tacos	$2,054.35	($86.96)	$1,967.39
Coffee + Donuts	($913.04)	$239.13	($673.91)

If we want to study the EV scenarios on a rainy day, we *multiply the payoff matrix by the probabilities for a rainy forecast (77%/23%).*

TABLE 30: EXPECTED REVENUE ON A SUNNY DAY.

EV on a Rainy Day Course of Action	Sun (b1)	Rain (b2)	Total
Ice Cream + Tacos	$508.06	($774.19)	($266.13)
Coffee + Donuts	($225.81)	$2,129.03	$1,903.23

If the suboptimal outcomes are not of interest, multiply the complete payoff matrix and the conditional probability matrix and get a table of optimal strategies.

TABLE 31: OPTIMAL STRATEGIES.

Optimal Strategies Table	States of Nature		
Course of Action	Sun	Rain	Total
Ice Cream + Tacos	$2,054.35	($86.96)	$1,967.39
Coffee + Donuts	($225.81)	$2,129.03	$1,903.23

So far, the analysis using Bayes is pretty clear. *If **rain** is predicted, bring coffee + donuts. If it's **sunny**, bring ice cream + tacos.* But what about a cloudy day?

Analyzing the value of better prediction using Bayes

Assuming that we act according to the weather information available, such as packing ice cream and tacos when sun is predicted, Bayes tells us the improved prediction potential is $547.50 compared to the season's EV.

TABLE 32: EXPECTED REVENUE CALCULATIONS ACCOUNTING FOR CLOUDY DAYS.

Decisions	Weighted State	Optimal EV	Cloudy Day EV
Sunny Day	69%	$1,967.39	$1,357.50
Rainy Day	31%	$1,903.23	$590.00
Total			**$1,947.50**
Impact of forecast information on EV Daily Revenue ($1400)			**$547.50**
Potential Improvement due to better forecasting			**$452.50**

Key takeaways

- Probability helps make better decisions

- Probability can give you a sense of what information is worth and how much it can contribute to improving an outcome,

- Bayes allows you to further use probabilities by allowing you to update your probabilities with new information, like the accuracy of the day's forecast.

In the next chapter, we will tie together all of the tools we have learned thus far, to help our friends a GreenGear structure and analyze another big decision that could change the direction of their firm.

TWELVE

SuperPower: Making Decisions Under Uncertainty

ccounting for uncertainty is no simple task, but that's why you are the superhero in this story. Book smarts are great, but street smarts are better when you are on the ground, tackling problems and making tough decisions with limited data. In this chapter, we will visit with our friends at GreenGear, who are facing a new business opportunity driven by adopting new environmental protection policies. We will work through the decision problem with Bob and the gang and help them make sense of an uncertain situation.

GreenGear is a manufacturer of specialized energy-saving systems. They have an industrial offering called *SuperStack* that allows warehouses and large installations to improve energy efficiency by 30% and a residential offering called *PowerPack* that allows the average homeowner to reduce their power consumption by 42%. In North

America, there are only several players in this market, and up till now, GreenGear has been able to sell 100% of its production in both markets.

However, with governments becoming more and more concerned with energy consumption and the impact on power grids, different policy actions are being taken to subsidize retrofitting homes and businesses with these innovative power-saving technologies. More specifically, the municipality of *Big Sitee*, the hometown of GreenGear, is considering passing a bill where all homeowners would be required to retrofit their homes with PowerPack-type devices, and the Federal government is considering a similar policy concerning SuperStack-like devices but for large manufacturing companies and Bitcoin miners.

GreenGear is in a once-in-a-lifetime situation where they could both increase their production and be guaranteed to sell everything they make by legal mandate. This means that GreenGear needs to lease new equipment to meet the demand.

Defining the initial problem frame

Company founder Bob calls everybody into a meeting to see what manufacturing strategy they should pick to make the most of the policy initiatives ahead of time. Because Bob, Dave, and Ralph the marketing manager can't agree on the right move, they call you, the local neighborhood decision superhero, back in for some good ole decision support.

If they make the wrong decision or wait for the bills to play out in legislation, they could end up being late to the game by six months due to the long lead times required to deliver the production equipment from overseas. That would be enough time for another player to swoop in or simply translate into lost sales.

We can start documenting our decision frame in a simple table stating what we understand about the decision. This table will be updated periodically to reflect our evolving state of knowledge on the decision to be made. *In bold are the elements that are considered the most important consideration in their respective column.*

TABLE 33: GREENGEAR'S INITIAL DECISION FRAME.

What we know	What we don't know	What matters
If a bill passes, we can sell 100% of our production for that product. Both bills can pass.	Which bills will pass.	Capitalize on market opportunity + Maximize revenue.
Cost and production specs for each option.	Which machine to lease ahead of time.	Limit downside risk. Leasing equipment and no bill passes could cripple cashflow for years.
		Preventing competitors like Blue Dream to dominate the market first.

Now that the basic problem is framed, let's organize some data to better understand the situation. *The analysis is available in the example files so you can follow along.*

Organizing the decision problem

Dave, the finance director, had already put together a table outlining the possible leasing options for Bob to discuss. Dave explains they can get dedicated machines for either PowerPack or SuperStack or a machine that can accomplish both production lines.

TABLE 34: GREENGEAR'S LEASING OPTIONS.

| | Machine leasing options | | |
	PowerPack	SuperStack	Multi-Purpose
Annual Capacity	25,000	4,000	29,000
Lease Cost	$3,000,000	$12,000,000	$14,000,000
Sell Price	$500.00	$4,500.00	*$1,051.72*

PowerPacks are much smaller and easier to produce, so annual production is about 25,000 and they sell for $500/home. The leasing cost for the new PowerPack machine is $3M. Conversely, the SuperStack system is much bigger, costing $12M, and can produce 4,000 units, which sell for $4,500. If GreenGear leases the multi-purpose machine, they can produce both offerings and would have an average unit cost of $1,052/unit for all 29,000 units.

Identifying alternatives to analyze

With this information, you start by helping your friends at GreenGear frame the problem correctly by listing all the potential scenarios. Though it may appear obvious, there are still eight potential decision strategies (the math is (2^3)), that could include visibly sub-optimal strategies, such as leasing both machines versus a multi-purpose that does both at a lesser acquisition cost.

TABLE 35: STRATEGY ALTERNATIVES TABLE (DOMINANT STRATEGIES IN BOLD).

	PowerPack	SuperStack	Multi-Purpose
Strategy 1	Lease	Lease	Lease
Strategy 2	Lease	Lease	-
Strategy 3	**Lease**	**-**	**-**
Strategy 4	Lease	-	Lease
Strategy 5	-	Lease	Lease
Strategy 6	**-**	**-**	**Lease**
Strategy 7	**-**	**Lease**	**-**
Strategy 8	**-**	**-**	**-**

Using Savage's rule on dominance, we know that not all of these options are useful in light of the others that are available in that strategy. For example, if you lease a multi-purpose machine, that negates the need to lease the other machines. Looking at the table, there are four dominant (or distinct) strategies to assess:

- Lease the PowerPack Machine (Strategy 3)
- Lease the SuperStack Machine (Strategy 7)
- Lease the Multi-Purpose Machine (Strategy 6)
- Lease nothing (differ to bill passing) (Strategy 8)

The other key dimensions to define are the states of nature impacting the strategies. As we mentioned earlier, two policy projects on the books could impact GreenGear, but neither has passed yet. This means we have *four states of nature* because **either**, **neither**, or **both bills** could pass and be voted into law.

TABLE 36: MAPPING COURSES OF ACTION TO STATES OF NATURE.

		S_1	S_2	S_3	S_4
			States of Nature		
	Product Strategy	Neither Bill	Municipal Program	Federal Program	Both Bills
COA_1	PowerPack	COA_1S_1	COA_1S_2	COA_1S_3	COA_1S_4
COA_2	SuperStack	COA_2S_1	COA_2S_2	COA_2S_3	COA_2S_4
COA_3	Multi-Purpose	COA_3S_1	COA_3S_2	COA_3S_3	COA_3S_4
COA_4	No Lease	COA_4S_1	COA_4S_2	COA_4S_3	COA_4S_4

We have our decision analysis table by organizing the courses of action (COA_i) and states of nature (S_i) into a 2x2 matrix. With four states of nature and four courses of action, we evaluate 16 different potential scenarios. This will serve as the framework for analyzing the decision with Bob and his team.

Analyzing the decision options

Using all the information GreenGear has provided, we will work through the decision problem using classic methods such as payoff tables and probability trees only. Meaning this analysis can be carried out using simple Excel or Google Sheets.

Decision rules apply to Monte-Carlo too!

The approach and how to apply the decision rules do not limit themselves to analyzing payoff tables. They are the foundational basis for analyzing the results from more powerful and sophisticated methods, such as Monte-Carlo simulation, to which we have dedicated two entire books!

Payoff analysis

For each of the 16 scenarios, we can assign a number of guaranteed units to sell. Bob mentioned earlier that GreenGear was in a position to sell 100% of its production, so with that in mind, we estimate the sales to be the same as the maximum capacity for each machine.

TABLE 37: DEMAND PROFILE ORGANIZED IN A DECISION MATRIX.

Demand Scenarios

Product Strategy	Neither Bill	Municipal Program	Federal Program	Both Bills
		States of Nature		
PowerPack	0	25,000	0	25,000
SuperStack	0	0	4,000	4,000
Multi-Purpose	0	25,000	4,000	25000/4000
No Lease	0	0	0	0

Now the fun begins! We will calculate the Payoffs for each scenario. The formula, in this case, is quite simple:

As you can imagine, this is easy to do in your favorite spreadsheet. Imagine running this workshop in the old days using a calculator or even a slide rule! We live in wonderful times, but I digress.

The payoff table tells us which options have the best maximum pay-out. Here, we look at which options have the best payout for a given strategy (COA), which is also known as the *maximax* decision rule.

TABLE 38: GREENGEAR'S PAYOFF MATRIX.

Payoff Matrix

Product Strategy	States of Nature			
	Neither Bill	Municipal Program	Federal Program	Both Bills
PowerPack	($3,000,000)	$9,500,000	($3,000,000)	$9,500,000
SuperStack	($12,000,000)	($12,000,000)	$6,000,000	$6,000,000
Multi-Purpose	($14,000,000)	($1,500,000)	$4,000,000	$16,500,000
No Lease	$0	$0	$0	$0

Ralph, Bob's marketing guy, looks at this and says, "Wow, this is great. We should get the multi-purpose machine. We will always make more!" prompting Dave to answer, "Of course, you would bet the farm on a chance of winning. Did you consider that maybe only the municipal bill passes? What happens if we lease the wrong machine or none of the bills pass? That could be $12-14M and that's disastrous! I would be way more comfortable going with the PowerPack because, at worst, we lose $3.0M, a number I find much more workable." Dave's risk-averse nature is clear because his preferred strategy is a *maximin*, meaning he prefers to minimize the worst outcome.

Another interesting metric is calculating a bias for optimism or pessimism, also known as the Hurwicz Criterion. It is a weighted average of the upper and lower bounds. If you are optimistic, it will be >0.5 and if you are pessimistic it will be < 0.5.

So you turn around and ask Ralph how he feels about the prospects, and he says he is pretty good and would set the criterion to 0.6. Using this decision rule, the PowerPack strategy comes up first again, prompting Dave to say, "See! PowerPack has a better chance of minimizing loss—just as I thought.."

TABLE 39: CALCULATING THE OPTIMISM COEFFICIENT (HURWICZ CRITERION).

Payoff Matrix

Product Strategy	States of Nature				Hurwicz Criterion (0.6)
	Neither Bill	Municipal Program	Federal Program	Both Bills	
PowerPack	($3,000,000)	$9,500,000	($3,000,000)	$9,500,000	$4,500,000
SuperStack	($12,000,000)	($12,000,000)	$6,000,000	$6,000,000	($1,200,000)
Multi-Purpose	($14,000,000)	($1,500,000)	$4,000,000	$16,500,000	$4,300,000
No Lease	$0	$0	$0	$0	$0

Bob chimes in with, "Gentlemen, let's give our friend some time to help us. I know we still have a ways to go in our assessment because every decision rule is telling us something different!"

Analyzing regrets

You reassure Bob and Dave by telling them, "Hold on guys, Bob's absolutely right. We are not done yet. We have to look at the other side of the equation. For this, we need to calculate the *maximum regret* between the chosen strategy and the worst outcome for that state of nature. Remember, regret is the difference between the payoff of the optimal decision and the payoff of the decision made. For example, to calculate the regret for $COA_1S_1 = (\$3M)-(\$14M) = \$11.0M$."

TABLE 40: GREENGEAR'S REGRET MATRIX.

Regret Matrix					Regret Analysis
	States of Nature				
Product Strategy	Neither Bill	Municipal Program	Federal Program	Both Bills	Minimax Regret
PowerPack	$11,000,000	$21,500,000	$0	$9,500,000	$21,500,000
SuperStack	$2,000,000	$0	$9,000,000	$6,000,000	$9,000,000
Multi-Purpose	$0	$10,500,000	$7,000,000	$16,500,000	$16,500,000
No Lease	$14,000,000	$12,000,000	$3,000,000	$0	$14,000,000

The maximum regret rule compares the maximum regret from all the strategies, recommending you take the lower number in the group. *A lower regret means that the course of action would have been in hindsight better than the other strategies on offer.* The lower the regret, the more attractive the option when using this decision criteria. This approach aims to minimize potential feelings of remorse from a decision. Bob looks at the regret table and says, "So, if we want to make sure we made the least painful decision, we should now be thinking about the SuperStack?" You answer, "Maybe, maybe not. We still need to factor in the bills passing or not."

Accounting for uncertainties using probability

If we take the naïve scenario where everything has an equal chance of occurring, that the smart people call Laplace's rule of equal probability, we can calculate an average payout per strategy and on average, PowerPack is a better strategy.

TABLE 41: EV OF PAYOFFS USING LAPLACE'S EQUAL PROBABILITY RULE.

Weighted Scenarios					
		Conditional probabilities of bills passing			
	25%	25%	25%	25%	
		States of Nature			
Product Strategy	Neither Bill	Municipal Program	Federal Program	Both Bills	Weighted Payoff
PowerPack	($750,000)	$2,375,000	($750,000)	$2,375,000	$3,250,000
SuperStack	($3,000,000)	($3,000,000)	$1,500,000	$1,500,000	($3,000,000)
Multi-Purpose	($3,500,000)	($375,000)	$1,000,000	$4,125,000	$1,250,000
No Lease	$0	$0	$0	$0	$0

"But wait, averages are bad, Bob, don't you have a friend lobbying for the federal program?" asked Dave. "Yeah, I spoke to him last week and he is pretty sure it's going to happen. He is almost 80% certain on that. I think Ralph might have a better idea what's going on at city hall, you know his sister is the mayor's assistant." "Very funny Bob, but yes, I did ask, and things aren't looking very good. If I had to guess, it's probably less than chance, I would say about 40%."

"Well gentlemen, that is very interesting. Let's work that into our analysis using Expected Value. But as you know, we can't use the probabilities as-is, for this we need to go Bayesian. First, we need to calculate the conditional probabilities. To do this, I set up a table to lay out the numbers, but before we begin, are these bills related in any way? Does one affect if the other is more likely to be adopted?" Bob answers confidently, "No buddy, my friend in the capital tells me they are completely unrelated." "Ok, Good. This will be simple."

TABLE 42: CONDITIONAL PROBABILITIES OF A BILL PASSING.

Probabilities of bills passing		
	Pass	Not Pass
Municipal Program	40%	60%
Federal Program	80%	20%

Since we have two outcomes and two independent ones, we can calculate the conditional probabilities by multiplying them together, as shown below. This allows us to assign specific conditional probabilities to each of the states of nature under study. Very important if we want to improve model fidelity. When interpreting the formulas, **T** means *True* and **F** means *False*.

Outcome	Calculating Conditionals	Conditional Probability
Neither pass	MUNICIPAL(F)*FEDERAL(F) =	0.6 x 0.2 = 12%
Municipal only	MUNICIPAL(T)*FEDERAL(F) =	0.4 x 0.2 = 8%
Federal only	MUNICIPAL(F)*FEDERAL(T) =	0.6 x 0.8 = 48%
Both pass	MUNICIPAL(T)*FEDERAL(T) =	0.4 x 0.8 = 32%

We get the following table by multiplying all the outcomes from the payoff matrix by the appropriate conditional probability:

TABLE 43: FACTORING IN CONDITIONAL PROBABILITIES TO CALCULATE EV.

Weighted Scenarios

	Conditional probabilities of bills passing				
	12%	8%	48%	32%	
	States of Nature				
Product Strategy	Neither Bill	Municipal Program	Federal Program	Both Bills	Weighted Payoff
PowerPack	($360,000)	$760,000	($1,440,000)	$3,040,000	$2,000,000
SuperStack	($1,440,000)	($960,000)	$2,880,000	$1,920,000	$2,400,000
Multi-Purpose	($1,680,000)	($120,000)	$1,920,000	$5,280,000	$5,400,000
No Lease	$0	$0	$0	$0	$0

Using the probabilities and conditional probabilities from GreenGear, we can put together a simple decision tree to aid in understanding the problem. The expected value for each option in the decision tree aligns with those in our weighted scenarios table.

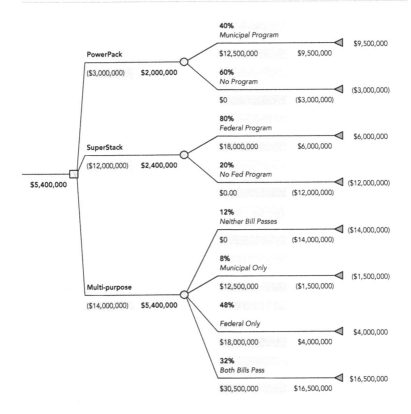

Figure 70:Probability tree for GreenGear's decision problem.

Calculating the utility of alternatives

You tell the team that you pulled the utility curve from the Blue Dream Manufacturing deal and adapted it to this decision, scaling the curve from the worst outcome (-$14M) and the best outcome ($16.5M). As we remember from Chapter 10, we can measure the utility of alternatives using a correctly calibrated utility function.

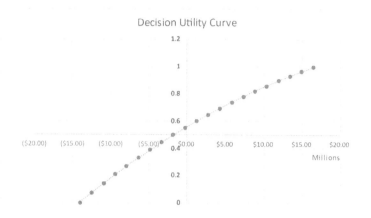

Figure 71: GreenGear's utility curve.

The utility of each option implied by the utility curve for each strategy is listed below. Utility and Weighted CME suggest that the multi-purpose machine is the best strategy for GreenGear, while PowerPack is recommended to use the CME.

TABLE 44: CALCULATING CME AND UTILITY METRICS.

Decision Rule	Utility	Economic CME	Weighted CME
PowerPack	0.62	$3,954,555	$474,547
SuperStack	0.64	($1,985,440)	($158,835)
Multi-Purpose	0.73	$2,969,115	$1,425,175
No Lease	0.56	$0	$0

What to do?! With so many rules (nine to be exact), Bob, Ralph, and Dave have good reason to still be arguing over the best decision. Let's help them out.

Comparing alternatives

To make things easier, we have summarized all the metrics into one table with the winning option highlighted in green.

TABLE 45: DECISION RULE COMPARISON TABLE.

Decision Analysis Table

Decision Rule	PowerPack	SuperStack	Multi-Purpose	No Lease
Scenario Probability	8%	48%	32%	12%
Min Payoff	($3,000,000)	($12,000,000)	($14,000,000)	$0
Max Payoff	$9,500,000	$6,000,000	$16,500,000	$0
Equal Probability	$3,250,000	($3,000,000)	$1,250,000	$0
Hurwicz Criterion (0.6)	$4,500,000	($1,200,000)	$4,300,000	$0
Minimax Regret	$21,500,000	$9,000,000	$16,500,000	$14,000,000
Weighted Payoff	$2,000,000	$2,400,000	$5,400,000	$0
Utility	0.62	0.64	0.73	0.56
Economic CME	$3,954,555	($1,985,440)	$2,969,115	$0
Weighted CME	$474,547	($158,835)	$1,425,175	$0

Now, with the numbers in hand, the conversation has become even livelier than before! Ralph suggests we should eliminate the *No lease* option because GreenGear has no potential for gain. Dave agrees. "You're right. You can't win if you don't play. Anyways, and for that matter, do we really want Jim at Blue Dream to come and eat our lunch? We need to win in any case, so I think the SuperStack should be dropped too. Let's face it, it's only an option because it's the least regrettable choice with the highest likelihood of occurring."

Bob turns to both of them and says, "Wait guys, you are acting like each green box is a vote and I don't think this is what is intended, right?" Relieved, you answer, "Exactly, Bob, you get it. What you really need to do is first decide what's important in the decision. Let me write a few down on the whiteboard."

Are we more concerned about?

- If a loss is our biggest worry, we should look at the smallest loss scenario (Maximin rule)

- If the size of the prize is the main consideration, then we should primarily consider the minimum payout (Maximax rule).

- If betting on the wrong horse keeps you up at night, analyze the decision based on minimizing regret.

- If you are risk neutral, then pick your best option using the expected value or weighted payout.

- Selecting the most likely scenario, use conditional probability

"Each rule has a set of assumptions and risk preferences associated with it. As a group, they need to weigh which ones are most important."

Making the decision

You tell Bob that the next step is to revisit the problem frame and use that to pick a set of rules that makes sense. Below is the framing/scoping table for the decision. Based on the team discussions, we noted the following and put in bold the team's most important item in the column to consider :

TABLE 46: UPDATED DECISION FRAME FOR SELECTING THE RIGHT PRODUC-
TION STRATEGY.

What we know	What we don't know	What matters
If a bill passes, we can sell 100% of our production for that product. Both bills can pass.	Which bills will pass.	Capitalize on market opportunity + Maximize revenue.
Cost and production specs for each option.	**Which machine to lease ahead of time**	**Limit downside risk. Leasing equipment and no bill passes could cripple cash-flow for years.**
There is a 88% of at least 1 bill passing.		Preventing competitors like Blue Dream to dominate the market first.

By agreeing and focusing on the elements of the decision, it becomes easier to pick the right rules to solve it. *At a fundamental level, the most important thing we don't know should be the base question the analysis intends to help resolve.* If not, perhaps it's time to slow things down and ensure we are trying to solve the right question.

One of the important pieces of information appears in the table (in bold): *there is an 88% chance of a bill passing!* So maybe that is a good reason to discard the no-lease option as Ralph suggested. But which machine to lease?

The PowerPack option is interesting financially because of its lower lease cost and revenue potential, if not because it has only an 8% chance of happening alone. Once again, *probability dictates* that maybe that's not a good place to bet the farm.

The SuperStack product line has a good profile, but 52% chance of that law is not passing, which would mean a $12M, a loss Dave does

NOT approve of, and the most important consideration in the *What matters* column in Table 46.

The last standing strategy is the *multi-purpose machine*. Because this machine makes both PowerPack and SuperStack, it is subject to the probabilities of four outcomes. By studying the sub-tree with all four states of nature (scenarios), we can gain a different perspective on the decision. First, there is a 12% chance of losing 14M, but that is five times better than SuperStack alone, where you risk 12M, 52% of the time. There is an 8% chance of losing 1.5M, though an unhappy outcome, it will not cripple the firm. 48% of the time you are making $4M and 32% of the time you are making $16.5M.

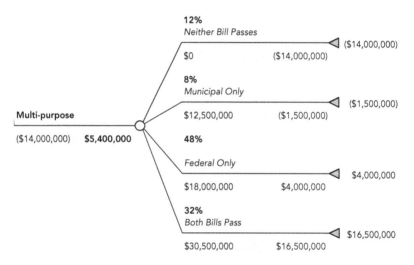

Figure 72: Multi-purpose machine probability tree.

Probability rules dictate that 80% of the time, you are making money with the multi-purpose machine and this option satisfies all the criteria laid forward in the frame. The analysis dictates that this is the best option, but this is a conversation and a decision that Bob and the gang will have to make on their own.

Key Takeaways

- We framed our problem using the three-what table to identify the important considerations in the GreenGear Case.

- We used a payoff table to analyze options, which told us which options were most attractive and we calculated which options had the least regret.

- We added probability to our decision model, mapped out a decision tree, giving different perspectives to the analysis and generating several supplementary decision metrics.

- We analyzed our options using nine different decision metrics. Then we used the decision frame to evaluate which options were most satisfactory.

THIRTEEN

Remembering the Important Stuff

As we have done for each chapter, we have collected a list of important takeaways for the entire book that every decision superhero can use.

- Being a decision superhero is not always easy. Being objective, tight deadlines and giving results people may not like.

- Keep an open mind and constantly seek new information to support ideas, hypotheses and recommendations.

- Always be learning. Being curious about current events, industry news and reading books like Decision Superhero!

- There is no magic structure of business configuration that will make decision science happen magically. Each organization is unique and whatever structure is employed, people have to communicate to be effective at making better decisions.

- Framing decisions correctly saves time. Lots of time.

- Humility. Recognizing that we do not know everything makes obtaining insight and sincere opinions easier from people.

- Almost anything can be quantified and measured so there are no excuses for not estimating.

- Decision models are represented using math, logic and/or diagram. Models are the base currency of decision science and their subtleties should be understood.

- Not all decisions merit the same analysis. There is always a threshold that defines when analyzing is no longer economically warranted. Some decisions can even be automated.

- Decision Scientists and Data Scientists can and should work together. These are complimentary roles that both thrive on improving outcomes.

- Always ask yourself if you want faster decisions, better decisions, or a combination of both. Figuring out what your objective is will dictate how to approach the problem.

- The best way to get people to like decision science is for you to show results early and often with quick wins. Use a Pareto analysis to allocate your efforts for the biggest bang for the buck. This is covered in book 2.

- All decisions have inputs, a frame, and objectives. Courses of action and states of nature provide a framework for structuring and making decisions.

- Probability is a powerful tool to model and rank decisions. From estimating the probability of rain to making decision trees, learning about probability will give you a new way to see the world.

- Understanding what kind of decision you are working on helps define the type of math and models you need. Decisions under risk, decisions under uncertainty, or decisions under conflict all merit different approaches.

- Some people love risk while others avoid it. Knowing who's who has a big impact on which and how recommendations are made.

- *Value of Perfect Information* can help the decision superhero to plan and budget for better data while keeping in mind what it is worth and how much is needed to make a difference.

FOURTEEN

Decision Superhero 2 and 3

By now, you know that a decision scientist is a fancy term for people who are good at making models for decision-making. The toolsets are usually a combination of methodology and math seeking to answer different business questions. A decision superhero should be able to identify where these tools are, gain access to them, conduct research in an empirical way, and translate their insights into a mathematical model that provides decision clarity that is obvious to others. Ultimately, the purpose of applying math is to have a greater understanding and a sound basis for making a decision by having a model that can entertain the expected uncertainties.

As a decision superhero, by reading this book, you have acquired the ideas, philosophical underpinnings, and foundations of Decision Science. Taking an intellectual approach is not always popular but always fruitful, especially when combined with hard mathematical skills and techniques. By hard, we do not mean difficult, but rather tangible.

Originally written as one book, the sheer vastness of the topic of decision science combined with practical examples in Julia and Microsoft Excel prompted us to split off the technical material into two other books.

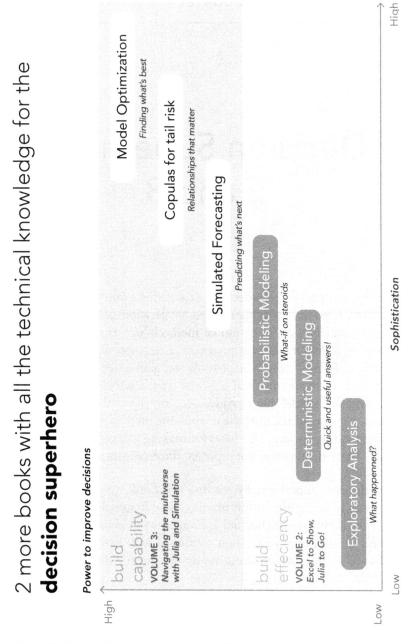

Figure 73: Decision Superhero Technical Skills Roadmap

About *Decision Superhero 2: Excel to show and Julia to go!*

Book 2 in the Decision Superhero series builds on the statistical decision theory methods we covered in Chapters 9-12 and expands the discussion to include computational thinking, logic, model development, what-if analysis, and deterministic and probabilistic simulations. All the examples are presented in Excel to explain the idea and algorithm and in Julia to show the code implementation of the same technique or model. The objective is to leverage your Excel modeling skills to learn how to automate models and speed up your time to answer using Julia. Prototype in Excel and develop in Julia—it's that simple. The concepts relating to programming are universal and can be applied in most other languages.

Some of the big themes and techniques covered include:

- Modeling and information presentation best practices

- How to set up your development environment to maximum effect

- Universal data patterns and programming concepts

- Developing Monte Carlo simulations using probability distributions to reflect sources of uncertainty

- Correlation and how to both implement it and analyze it to see which variables are driving the output of interest.

- Distribution fitting and how to pick the right distribution to best reflect subject matter opinions.

- Using probability models to predict outcomes (Markov Chains, Martigales).

- Estimating variables using analytical methods (Learning Curves, Pareto Analysis, Exponential Growth Models, Curve Fitting).

About *Decision Superhero 3: Navigating the multiverse using simulation and Julia*

Book 3 is when you are ready to take on some serious modeling, focusing on advanced modeling techniques such as fitting forecasting methods using historical data, forecast simulation and stress testing, numeric optimization, and copulas for advanced correlation. We learn how to build sophisticated business models such as discounted cashflows, simulate and optimize portfolios, and make all of it run blazing fast using Julia's native multi-threading and parallel computing facilities.

Different types of analytical approaches require different levels of sophistication to achieve results. Usually, the starting point for almost any organization is to look at historical performance using reports or interactive dashboards, otherwise known as business intelligence. The next natural step is to anchor yourself or, rather, the basis for the decision in a model.

Forecasting has the objective of giving you a clear enough perspective on the future to allow you to decide but has the limitation of being a deterministic best guess in most conventional settings. Another way is to build a probabilistic simulation model, which gives you a probability of success based on many potential outcomes rather than the best guess. Simulations use brute force to test out thousands of randomly generated scenarios and allow you to empirically calculate the probability of success. The interesting similarity between conventional time-series analysis and probabilistic modeling is that they share the same absolute goal—providing the basis or evidence for making a decision.

If you combine time-series forecasting and simulation, you could stress-test your model over time and get a probability of success. The goal is far less predicting the future than the ability to see whether you can survive all the uncertain conditions over a given period in the future. Lastly, it may be desirable to find out what is best by attempting to see which combination of variables under the decision-maker's control will lead to the best outcome.

Obviously, certain methods yield more interesting and useful answers because they incorporate uncertainty and answer more sophisticated questions. Mobility is not linear, and there is no reason why an organization cannot move from a business intelligence perspective to one that incorporates simulation and simulated forecasts to assign the probability of success to its various business decisions.

Part 2 of the book will review these approaches to answering business questions and provide practical models and examples in Julia and Excel for simulation, forecasting, and optimization.

Bibliography

Ackoff, R. L. (1956). The Development of Operations Research as a Science. *Operations Research, 4*(3), 265-295.

Aristotle. (340 BC). Teachings. Stagira: Self Published.

Ashkenas, R., Ulrich, D., & Kerr, S. (1995). The Boundaryless Organization. San Francisco: Josey Bass.

Barney, J. (1991). Firm Resources and Sustained Competitive Advantage. Journal of Management, Vol.17 No. 1.

Bergstrom, C. T., & West, J. D. (2020). Calling bullshit : the art of skepticism in a data-driven world. New York: Random House.

Bernoulli, D. (1954). Exposition of a New Theory on the Measurement of Risk. Econometrica, 22(1), 23-36.

Bernstein, P. L. (1996). Against the Gods: The Remarkable Story of Risk. New York: John Wiley and Sons.

Binder, J., & Watkins, M. D. (2024, January-February). To Solve a Tough Problem, Reframe It. Harvard Business Review, 80-89.

Bloch, A. (1985). Murphy's Law Complete Edition : All the reasons why everything goes wrong. London, UK: Mandarin Press.

Boston Consulting Group. (2020, June 22). The Most Innovative Companies Ranking over Time. Retrieved December 30, 2020, from https://www.bcg.com/publications/2020/most-innovative-companies/data-overview

Brown, V. J. (2014, Oct. 1). Risk Perception: It's Personal. Environmental Health Perspectives, A276–A279. Retrieved from US National Library of Medicine: https://www.ncbi.nlm.nih.gov/pmc/articles/PMC4181910/

Budiansky, S. (2013, March 29). Blackett's War: The Men Who Defeated the Nazi U-Boats and Brought Science to the Art of Warfare. (The Washington Post) Retrieved December 8, 2020, from https://www.washingtonpost.com/opinion/

blacketts-war-the-men-who-defeated-the-nazi-u-boats-and-brought-science-to-the-art-of-warfare-by-stephen-budiansky/2013/03/29/3083879a-75ee-11e2-8f84-3e4b513b1a13_story.html

Center for Health Decision Science, Harvard T. H. Chan School of Public Health. (n.d.). What is Decision Science? (Center for Health Decision Science, Harvard T. H. Chan School of Public Health) Retrieved December 8, 2020, from https://chds.hsph.harvard.edu/approaches/what-is-decision-science/

Chernih, A., Maj, M., & Vanduffel, S. (2007, April). Beyond Correlations: The Use and Abuse of Copulas in Economic Capital Calculations. Belgian Actuarial Bulletin, 7(1), 0-0. Retrieved from http://ssrn.com/abstract=1359578

Chladek, N. (2018, July 26). GROUP DECISION-MAKING TECHNIQUES FOR SUCCESS. Retrieved from Harvard Business School Online: https://online.hbs.edu/blog/post/group-decision-making-techniques

Christian, B., & Griffiths, T. (2017). Algorithms to live by: The computer science of human decisions. New York, NY: Picador.

Clemen, R. T., & Reilley, T. (1999, February). Correlations and Copulas for Decision and Risk Analysis. Management Science, 45(2), 208-224.

Cooper, D. R., & Schindler, P. S. (2003). Business Research Methods. New York: McGraw-Hill.

Couillard, Guy (2001), Change Agents Skills Workshop, Montreal, Organizational Transition Associates Inc.

Crockett, Z. (2016, August 2). The Time Everyone "Corrected" the World's Smartest Woman. (Priceonomics) Retrieved December 14, 2020, from https://priceonomics.com/the-time-everyone-corrected-the-worlds-smartest/

Davenport, T. H., & Harris, J. G. (2007). Competing on Aalytics: The New Science of Winning. Boston, MA: Harvard Business School Press.

Dickson, R. C. (2009). The father of Guestology: An interview with Bruce Laval. The Journal of Applied Management & Entrepreneurship, 12(2), 80-99.

Dorey, M., & Joubert, P. (2007). Modeling Copulas: An Overview. The Staple Inn Actuarial Society.

Ezell, G. S. (2017). Organizational Decision Processes. In Improving Homeland Security Decisions. Cambridge University Press.

Fagin, B. (2014). Lava Lamp Random Number Generator. Retrieved December 29, 2020, from https://gist.github.com/UnquietCode/99b69b99e00ce52e34c1

Fisher, R. (2000). Numbers are Essentiaal: Victory in the North Atlantic Reconsidered, March -- May 1943. (familyheritage.ca) Retrieved December 14, 2020, from http://www.familyheritage.ca/Articles/victory1943.html

Fortune. (2020, 9 15). Fortune 500: Walmart. (Fortune Media IP Limited) Retrieved 12 24, 2020, from https://fortune.com/company/walmart/fortune500/

Gregory S. Parnell, T. A. (2013). Handbook of Decision Analysis. Hoboken: Wiley.

Grossman, R. L. (2014). Organizational Models for Big Data And Analytics. Journal of Organizational Design, 3(1), 20-25.

Hagel, J. (2002). Out of the box: Strategies for achieving profits today and growth tomorrow through Web Services. Boston, MA, USA: Harvard Business Press.

Harris, R. (2017, 07 28). Excel errors: How Microsoft's spreadsheet may be hazardous to your health. Retrieved from zdnet.com: https://www.zdnet.com/article/excel-errors-microsofts-spreadsheet-may-be-hazardous-to-your-health/

Hopkins, A. (2009). Failure to Learn: the BP Texas City Refinery Disaster. Sydney: CCH Australia.

Howard, R. A., & Abbas, A. E. (2016). Foundations of Decision Analysis. London, UK: Pearson Education.

Hubbard, D. (2009). How To Measure Anything. New York: John Wiley and Sons.

Husna, N. (2015). Introduction OF OPERATIONS RESEARCH. Retrieved from prezi.com: https://prezi.com/-mgruap5uvcr/introduction-of-operation-research/

Iman, R. L., & Conover, W. J. (1982). A distribution free approach to inducing Rank Correlation among input variables. Communications in Statistics - Simulation and Computation , 11(3).

INFORMS. (n.d.). Patrick M. S. Blackett. Retrieved December 8, 2020, from https://www.informs.org/Explore/History-of-O.R.-Excellence/Biographical-Profiles/Blackett-Patrick-M.-S

Institute of Industrial and Systems Engineers. (2020). Industrial and Systems Engineering Body of Knowledge. Norcross, Georgia: Institute of Industrial and Systems Engineers.

Keeney, R. (1996). Value-Focused Thinking: A Path to Creative Decisionmaking. Cambridge, Massachusetts: Harvard University Press.

Keeney, R. (2020). Give Yourself a Nudge. Cambridge: Cambridge University Press.

Keeney, R. L. (2004). Making Better Decision Makers. Decision Analysis, 1(4), 193-204.

Keeney, R. R. (1992). Value-Focused Thinking. Cambridge, MA: Harvard University Press.

Kernan, S. (2020, August 16). The Allies Hired a Brilliant Mathematician and It Paif Off, Bigtime. (Mdium.Com) Retrieved December 8, 2020, from https://medium.com/history-of-yesterday/how-abraham-wald-saved-allied-planes-by-asking-the-right-questions-de972c7282ea

Kirby, M. W. (2003). Operational Research in War and Peace. London: Imperial College Press.

Kolmogorov, A. N. (1956). Foundations of the theory of probability. New York: Chealsea Publishing .

Kouatchou, J. (2018, 02 20). Basic Comparison of Python, Julia, R, Matlab and IDL. Retrieved from NASA Modeling Guru: https://modelingguru.nasa.gov/docs/DOC-2625

Landry, L. (2017, November 9). THE IMPORTANCE OF CREATIVITY IN BUSINESS. Retrieved December 30, 2020, from https://www.northeastern.edu/graduate/blog/creativity-importance-in-business/

L'Écuyer, P. (2017). HISTORY OF UNIFORM RANDOM NUMBER GENERATION. Proceedings of the 2017 Winter Simulation Conference, (p. 28).

Leibacher, H. (2020, May 5). Main Street USA Tribute Windows: Family and Executives. (World Of Walt) Retrieved December 14, 2020, from https://www.worldofwalt.com/main-street-usa-tribute-windows-family-and-executives.html

Los Alamos National Laboratory. (2012, February 3). JohnvonNeumann-LosAlamos. (Los Alamos National Laboratory) Retrieved December 14, 2020, from https://en.wikipedia.org/wiki/File:JohnvonNeumann-LosAlamos.gif

Makrdakis, S., Hogarth, R., & Gaba, A. (2009). Dance with Chance: Making Luck Work for You. Oneworld Publications.

Marcia W. Blenko, P. R. (2011, April 15). Create a decision-focused culture. Retrieved December 27, 2020, from https://www.bain.com/insights/decision-insights-7-create-a-decision-focused-culture/

Maslow, A. (1966). The Psychology of Science.

McNeilly, M. (2012). Sun Tzu and the Art of Business: Six Strategic Principles for Managers, Revised Edition. New York: Oxford Press.

McNitt, L. L. (1983). Basic Computer Simulation. Blue Ridge Summit, , PA, USA: TAB Books.

Merriam-Webster. (n.d.). Systems Analysis Definition. Retrieved 04 29, 2020, from Merriam-Webster.com dictionary: https://www.merriam-webster.com/dictionary/systems%20analysis

Microsoft Corporation. (2018). Future Computed: Artificial Intelligence and its role in society. Microsoft Corporation.

Miranda, G. M.-L. (2018, October 18). Building an effective analytics organization. Retrieved December 28, 2020, from https://www.mckinsey.com/industries/financial-services/our-insights/building-an-effective-analytics-organization

Morgenstern, J. v. (1953). Theory of Games and Economic Behavior . Princeton: Princeton University Press.

Nelsen, R. (2002). Properties and Applications of Copulas: A Brief Survey. Lewis and Clark College / Mount Holyoke College.

Nohria, N., Joyce, W., & Bruce, R. (2003). What really works? Harvard Business Review, Vol. 80., No. 7 pp 42-52.

Poole, D., & Raftery, A. E. (2000). Inference for Deterministic Simulation Models: The Bayesian Melding Approach. Journal of the American Statistical Association, 95(452), 1244-1255.

Reilly, R. T. (2001). Making Hard Decisions with DecisionTools. Pacific Grove, CA: Brooks/Cole.

Richardson, S. (1968). Operations Research for Management Decisions. New York: Ronald Press.

Richardson, S. B. (1965). Statistical Analysis. New York: Ronald Press.

Robinson, J. (1949). On the Hamiltonian Game (A Traveling Salesman Problem). Santa onica, California: The Rand Corporation.

Rogers, S., & Thompson, J. K. (2017). Analytics: How to win with intelligence. Technics Publications.

Roser, M. (2019, May). Moore's Law Transistor Count 1971-2018. Retrieved December 14, 2020, from https://en.wikipedia.org/wiki/File:Moore%27s_Law_Transistor_Count_1971-2018.png

Rowe, A. J. (2006). Decision Making: 5 Steps to Better Results. Boston, MA: Harvard Business Press.

Saipol Bari Abd-Karim, D. A. (2014). Managing Conflicts in Joint Venture Projects. International Journal of Property Science, 4(1), 1-15.

Samuel D. Bond, K. A. (2010). Improving the Generation of Decision Objectives. Decision Analysis, 7(3), 235-326.

Savage, L. J. (1971). The Foundations of Statistics, Second Edition. Garden City, New York: Dover Publications.

Savage, S. L. (2009). The Flaw of Averages. New York: Wiley & Sons.

Senge, P. M. (1990). The Fith Discipline. New York, NY: Doubleday.

Shannon, C. E. (1948). A Mathematical Theory of Communication. Bell Systems Technical Journal.

Silver, N. (2012). The Signal and the Noise: Why Most Predictions Fail - But Some Don't. Penguin Group.

Sloan, A. P. (1964). My Years at General Motors. Garden City, NJ: Doubleday.

Society of Decision Professionals. (2020). THE RAIFFA-HOWARD AWARD. Retrieved December 30, 2020, from https://www.decisionprofessionals.com/assets/images-rha/SDP_RH_Award_Info.pdf

Society of Decision Professionals. (n.d.). Raiffa-Howard Award Assessment of Organizational Decision Quality. (Society of Decision Professionals) Retrieved December 30, 2020, from https://www.decisionexcellence.org/

Society of Decision Professionals. (n.d.). SDP's Professional Code. Retrieved December 28, 2020, from https://www.decisionprofessionals.com/about/professional-code

Socrates. (430 BC). Musings. Ahtens: Self Published.

Spetzler, C. H. (2016). Decision Quality. Hoboken: Wiley.

Springer, C. H., Herlihy, R. E., Mall, R. T., & Beggs, R. I. (1968). Probabilistic Models: Volume Four of the Mathematics for Management Series. Homewood, Illinois: Rischard D. Irwin Inc.

Springer, C. H., Herlihy, R. E., Mall, R. T., & Beggs, R. I. (1968). Statistical Inference : Volume 2 of Mathematics for Management Series. Irwin Press.

Stubbs, E. (2011). The Value of Business Anakytics: Identifying the Path to Profitability. New York: John Wiley and Sons.

Subcommittee, I. O. (2019, April). Building Successful O.R. and Analytics Teams. Retrieved December 29, 2020, from https://www.informs.org/Explore/Building-Successful-O.R.-and-Analytics-Teams

The Causality Actuarial Society. (2002, 10 31). Correlation: Copulas and Conditioning. Retrieved from The Causality Actuarial Society: https://www.casact.org/sites/default/files/old/02pcas_venter.pdf

Tjan, A. (2001, February). Finally, a Way to Put Your Internet Portfolio in Order. Harvard Business Review, 76-85.

Torkia, E. (2019, 01 17). "To code or not to code, that IS the question! [for analytics strategy planners]. Retrieved from LinkedIn: https://www.linkedin.com/pulse/code-question-analytics-strategy-planners-eric-torkia-masc/

Torkia, E. R., & Cassivi, L. (2009). E-Collaboration: A Dynamic Enterprise Model. Hershey, PA, USA: IGI Global.

Ulrich, D., Ashkenas, R., & Kerr, S. (2002). The GE Work-Out. New York , NY: McGraw-Hill.

Unbekannt. (2017, May 30). Bernoulli, Daniel (1700-1782)-Portrait. (Wikipedia) Retrieved December 14, 2020, from https://en.wikipedia.org/wiki/File:ETH-BIB-Bernoulli,_Daniel_(1700-1782)-Portrait-Portr_10971.tif_(cropped).jpg

University of Waterloo. (n.d.). Optimal Road Trips: History of the TSP with road distances. Retrieved December 9, 2020, from http://www.math.uwaterloo.ca/tsp/us/history.html

Venkatraman, N. (1994). IT-Enabled Business Transformation: From Automation to Business Scope Redefinition. Sloan Management Review, 73-87.

Venkatraman, N., & Henderson, J. C. (1993). Strategic alignment: Leveraging information technology for transforming organizations. IBM Systems Journal.

Vose, D. (2008). Risk Analysis: A Quantitative Guide, 3rd Edition. New York: John Wiley and Sons.

Walmart. (2020). About Us. (Walmart, Inc.) Retrieved 12 24, 2020, from https://corporate.walmart.com/our-story

Wetherbe, J. C. (1979). Systems analysis for computer-based information systems. St. Paul: West Publishing Company.

Wikipedia. (2018, April 28). Decision quality. Retrieved December 28, 2020, from https://en.wikipedia.org/wiki/Decision_quality

Wikipedia. (2020, February 13). Stigler Diet. Retrieved December 8, 2020, from https://en.wikipedia.org/wiki/Stigler_diet

Wikipedia Contributors. (2020, 04 23). "Scientific method," Wikipedia, The Free Encyclopedia. Retrieved 04 24, 2020, from https://en.wikipedia.org/w/index.php?title=Scientific_method&oldid=952581714

Wikipedia contributors. (2020, 5 7). Read–eval–print loop. Retrieved 5 11, 2020, from https://en.wikipedia.org/w/index.php?title=Read%E2%80%93eval%E2%80%93print_loop&oldid=955378943

Wikipedia. (2020, December 5). All models are wrong. Retrieved December 29, 2020, from https://en.wikipedia.org/wiki/All_models_are_wrong

WillMcC. (2009, April 26). Monorail Coral. (Wikipedia) Retrieved December 14, 2020, from https://en.wikipedia.org/wiki/File:Monorail_Coral.jpg

Yate, M. (2017). The Ultimate Job Search Guide. New York: Adams Media.

Yau, N. (2011). Visualize This. Ney York: John Wiley and Sons.

Index

Symbols

A

B

C

D

data science xxxiii, 29, 42, 44, 66, 69, 81, 85, 150, 165
DEA model 56, 61
decision analysis xxxi, 5, 6, 7, 44, 54, 152, 217, 221, 231, 248, 269
decision analyst 25, 26
decision components 180
decision quality xxxi, 9, 27, 32
decision science
  explanation xxxii
  explanation of 5, 9
  origin of 2
decision science explanation xxxii, 5, 9
decision science origin 2
decision scientist xxxiii, 26, 29, 35, 48, 62, 63, 71, 73, 74, 77
decision superhero
  definition of a 24
  mantra for 30
  process to become a 23
  roles of a 27
  skills of xxxii
decision superhero definition 24
decision superhero mantra 30
decision superhero process 23
decision superhero roles 27
decision superhero skills xxxii
DES. *See* Discrete Event Simulation
descriptive model 125
dialectical inquiry 184
disciplined framing process 188
Discrete Event Simulation 121
domain expertise 41
domain knowledge 35, 176, 193
Duval, Robert 36
Dynamic Enterprise Alignment model xxxiii, 48

E

economic modeler 28
economic modeling 120
economics 7, 111, 130, 133, 143, 144, 227
elicitation vii, 1, 32, 39, 190
elicitor 76

DOWNLOAD RESOURCES AND FILES AT
https://decisionsuperhero.com

www.ingramcontent.com/pod-product-compliance
Lightning Source LLC
Chambersburg PA
CBHW071232050326

40690CB00011B/2083